UNIV. OF CALIFORNIA
WITHDRAWN
Chemistry Library

Macromolecular Symposia 211

QD
1
M332
v.211
CHEM

9th Dresden Polymer Discussion: Polyelectrolytes

Dresden, Germany
March 16–19, 2003

Symposium Editor:
U. Scheler, Dresden, Germany

pp. 1–233 · April 2004
ISBN 3-527-31044-4

Macromolecular Symposia publishes lectures given at international symposia and is issued irregularly, with normally 14 volumes published per year. For each symposium volume, an Editor is appointed. The articles are peer-reviewed. The journal is produced by photo-offset lithography directly from the authors' typescripts.
Further information for authors can be found at http://www.ms-journal.de
Suggestions or proposals for conferences or symposia to be covered in this series should also be sent to the Editorial office (E-mail: macro-symp@wiley-vch.de).

Editor: Ingrid Meisel
Senior Associate Editor: Stefan Spiegel
Associate Editor: Alexandra Carrick
Assistant Editors: Sandra Kalveram,
Mara Staffilani

Executive Advisory Board:
M. Antonietti (Golm), M. Ballauff (Bayreuth), S. Kobayashi (Kyoto), K. Kremer (Mainz), T. P. Lodge (Minneapolis), H. E. H. Meijer (Eindhoven), R. Mülhaupt (Freiburg), A. D. Schlüter (Berlin), J. B. P. Soares (Waterloo), H. W. Spiess (Mainz), G. Wegner (Mainz)

Macromolecular Symposia:
Annual subscription rates 2004
Macromolecular Full Package: including Macromolecular Chemistry & Physics (18 issues), Macromolecular Rapid Communications (24), Macromolecular Bioscience (12), Macromolecular Theory & Simulations (9), Macromolecular Materials and Engineering (12), Macromolecular Symposia (14):

Europe	Euro	6.424 / 7.067
Switzerland	Sfr	11.534 / 12.688
All other areas	US$	7.948 / 8.743

print only **or** electronic only / print **and** electronic

Postage and handling charges included. All Wiley-VCH prices are exclusive of VAT. Prices are subject to change.

Single issues and back copies are available. Please ask for details at: service@wiley-vch.de

Orders may be placed through your bookseller or directly at the publishers:
WILEY-VCH Verlag GmbH & Co. KGaA, P. O. Box 10 11 61, 69451 Weinheim, Germany, Tel. +49 (0) 62 01/6 06-400, Fax +49 (0) 62 01/60 61 84. E-mail: service@wiley-vch.de

For USA and Canada: Macromolecular Symposia (ISSN 1022-1360) is published with 14 volumes per year by WILEY-VCH Verlag GmbH Co. KGaA, Boschstr. 12, 69451 Weinheim, Germany. Air freight and mailing in the USA by Publications Expediting Inc., 200 Meacham Ave., Elmont, NY 11003, USA. Application to mail at Periodicals Postage rate is pending at Jamaica, NY 11431, USA. POSTMASTER please send address changes to: Macromolecular Symposia, c/o Wiley-VCH, III River Street, Hoboken, NJ 07030, USA.

© WILEY-VCH Verlag GmbH & Co. KGaA, Weinheim, Germany, 2004
Printing: Strauss Offsetdruck, Mörlenbach. Binding: J. Schäffer, Grünstadt

Macromolecular Symposia

Articles published on the web will appear several weeks before the print edition. They are available through:

www.ms-journal.de

www.interscience.wiley.com

9th Dresden Polymer Discussion: Polyelectrolytes
Dresden (Germany), 2003

Preface
U. Scheler

Author Index

Participants at the 9th Dresden Polymer Discussion: Polyelectrolytes conference.

Preface

The volume contains a selection of the results presented at the 9th Dresden Polymer Discussion 'Polyelectrolytes' held at the Evangelische Akademie in Meissen, Germany, March 16th to 19th 2003.

The 9th Dresden Polymer Discussion took place from March 16th to 19th at the Evangelische Akademie in Meissen. The Evangelische Akademie, an institution of the Evangelic Church in Saxony, was the host for the meeting for the fifth time. The nicely restored former monastery provided the accommodation and conference venue, and the interesting and close setting promoted informal discussion. Relaxation was provided by a chamber music concerto with Russian and Yiddish songs. A sightseeing tour led to the centre of the vine growing area with a visit to the nicely refurbished Schloss Wackerbarth. The tour included insights into the traditional and modern ways of wine making. Following a wine tasting there was a rustic dinner in the cellar of Schloss Wackerbarth.

The Dresden Polymer Discussion is a biannual meeting organized by the Institute of Polymer Research Dresden and the Institute of Macromolecular and Textile Chemistry of the Technische Universität Dresden on a selected topic from the area of Polymer Research. The 9th Polymer Discussion in 2003 was devoted to 'Polyelectrolytes'.

Polyelectrolytes are of wide practical application for the tailoring of surface properties like charge and hydrophobicity. There are widespread applications in water treatment and suspensions. Proteins and DNA exhibit behaviour of polyelectrolytes as well.

Participants came from Germany, Sweden, Russia, the Netherlands, France and the United States. The scientific program consisted of 29 invited lectures and 15 posters.

The aim of the ninth meeting of the series was to cover the aspects of polyelectrolytes from the synthesis, characterization and theoretical description to aspects of their application. Subjects of the investigation are polyelectrolytes in solution, complexes with polyelectrolyte and polyelectrolytes adsorbed on solid surfaces. Materials with dedicated properties such as stiff polyelectrolytes are synthesised. Advancements in the theoretical description and simulations provide further insight and assist the interpretation of experimental data.

The meeting was made possible by financial support from the German Science Foundation (DFG), the Saxon State Ministry for Science and the Fine Arts (SMWK), and the Institute of Polymer Research, and it was also sponsored by BASF AG, Degussa AG, and DSM Research. This continuous support for Dresden Polymer Discussion is gratefully acknowledged.

This symposium was a great scientific success because of the excellent and inspiring lectures. It certainly will be remembered amongst the participants for its stimulating discussions, which revealed that – in spite of the wealth of results – a large number of open questions remain in the field of polyelectrolytes, which will be addressed in the coming years.

The 10th Dresden Polymer Discussion shall be held in spring 2005 on Characterization of Polymer Surfaces and Thin Films.

I wish to thank all participants who contributed to the success of the 9th Dresden Polymer Discussion 'Polyelectrolytes' and, in particular, all those who contributed to this Macromolecular Symposia volume.

U. Scheler

Macromol. Symp. **2004**, *211*, 1-24

Synthesis and Properties in Solution of Rodlike Polyelectrolytes

Matthias Ballauff,[1] *Jürgen Blaul,*[1] *Birgit Guilleaume,*[1] *Matthias Rehahn,* *[2]
Steffen Traser,*[2] *Matthias Wittemann,*[1] *Patrick Wittmeyer*[2]

[1] Polymer Institute, University of Karlsruhe, Kaiserstrasse 12, D-76128 Karlsruhe, Germany
[2] Ernst-Berl-Institute for Chemical Engineering and Macromolecular Science, Darmstadt University of Technology, Petersenstrasse 22, D-64287 Darmstadt, Germany
E-mail: mrehahn@dki.tu-darmstadt.de

Summary: Efficient synthetic strategies are described for the preparation of rodlike polyelectrolytes based on the intrinsically rigid poly(*p*-phenylene). Uncharged precursors were first prepared via the Suzuki coupling and then characterized by different methods of polymer analysis. Finally, they were transformed into polyelectrolytes using macromolecular substitution reactions. Depending on the substitution pattern, the obtained polyelectrolytes are either soluble or insoluble in water. Using water-soluble derivatives, the Poisson-Boltzmann cell model was tested by osmotic measurements and small-angle X-ray scattering. It is shown that the cell model provides a good first approximation of the distribution of the counterions around the macroion but still underestimates their correlation. Moreover, the PPP polyelectrolytes show a very pronounced polyelectrolyte effect. Since the rodlike PPPs are very rigid in shape, this observation proves that the polyelectrolyte effect is caused by long-range intermolecular electrostatic repulsion of the dissolved macroions rather than due to conformational changes.

Keywords: cell model; counterion condensation; osmotic coefficient; polyelectrolytes; rodlike polymers

Introduction

Polyelectrolytes represent key compounds in living organisms as well as in materials and processes of our daily life. They especially show their benefits when dissolved in water. Thus, for tailor-making polyelectrolytes, profound understanding of their behavior in aqueous solutions is required. It is well-known that electrostatic forces, osmotic effects and conformational changes must be considered in detail for developing this knowledge. Consequently, much effort has been spent in analyzing this class of functional polymers, using different techniques and materials. However, a quantitative interpretation of the collected data has been impossible up to today. This is because the intramolecular and intermolecular coulomb forces — and hence all properties

© 2004 WILEY-VCH Verlag GmbH & KGaA, Weinheim   DOI: 10.1002/masy.200450701

influenced by electrostatic interactions — strongly depend on the ionic strength: at very low ionic strength, polyelectrolyte molecules repel each other over very long distances. Moreover, the electrostatic forces cause intramolecular repulsion of the chain segments. Stretching and coil expansion are consequences in the case of flexible chains. These conformational changes, however, again change the intermolecular interactions. To conclude, it is impossible to differentiate the macroscopically measured data quantitatively according to the underlying intermolecular and intramolecular electrostatic, conformational and osmotic effects without profound theoretical understanding of polyelectrolyte behavior. Conversely, however, it is impossible to develop the needed knowledge by only considering flexible polyelectrolytes. Rigid, rodlike polyelectrolytes, on the other hand, cannot change their shape significantly. Therefore they represent ideal model systems for developing the required understanding. Poly(p-phenylene) (PPP) seems to be the most appropriate polymer backbone here because (*i*) it is intrinsically rodlike and (*ii*) it is inert to hydrolysis and other side reactions. Hence we developed powerful synthetic routes for the preparation of PPP-based polyelectrolytes. We combined the Pd-catalyzed Suzuki reaction with the concept of solubilizing side chains, and we applied macromolecular substitution routes (route B in Scheme 1) rather than direct syntheses according to route A. This procedure was selected because in precursor strategies the polycondensation process (B1) leads to a non-ionic intermediate which can be characterized using all techniques of polymer analysis. Provided an efficient substitution reaction is available for step B2, the well-defined precursors can subsequently be converted into the desired PPP polyelectrolytes while all molecular information determined for the uncharged precursor remains valid.

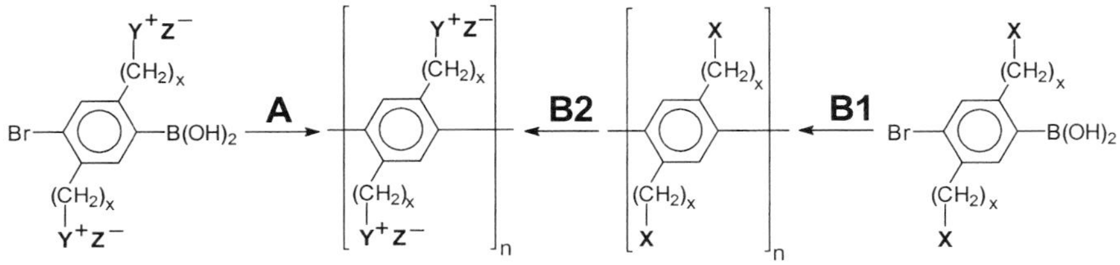

Scheme 1:   Synthetic approaches to PPP-based polyelectrolytes: direct synthesis (**A**) and precursor route (**B**); X: precursor functionality, Y^+: polyelectrolyte functionality, Z^-: counterion

© 2004 WILEY-VCH Verlag GmbH & KGaA, Weinheim

Synthesis of PPP Polyelectrolytes via Ether Intermediates

Aliphatic ethers were selected as precursor functionalities because they are inert under the Suzuki conditions but can be converted easily and completely into, for example, alkyl halogenides. In turn, these lateral alkyl halogenides should allow efficient introduction of ionic functionalities into the PPP in the final step.

In a first set of experiments, butoxymethylene-substituted precursor PPPs **3** were prepared.[1,2] The polycondensation of equimolar amounts of **1** and **2** leads to constitutionally homogeneous products **3** having values of $P_n \approx 60$ (Scheme 2). The lateral benzyl-alkyl ether groups were then cleaved quantitatively, leading to readily soluble bromomethylene-functionalized precursors **4** which were finally converted into carboxylated PPPs **5**, for example.

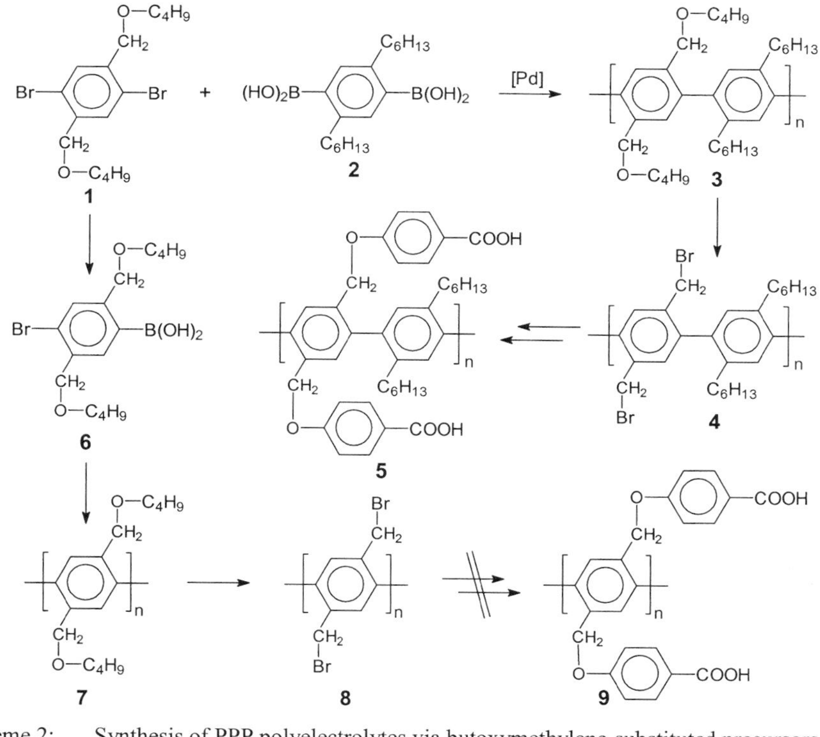

Scheme 2:   Synthesis of PPP polyelectrolytes via butoxymethylene-substituted precursors

© 2004 WILEY-VCH Verlag GmbH & KGaA, Weinheim

However, **5** proved to be insoluble in water or aqueous bases. We assumed that this is due to its low density of charged groups along the chains. Moreover, the apolar alkyl side chains attached to every second phenylene moiety might cause intermolecular hydrophobic interactions, leading to the observed insolubility in water. We therefore tried to prepare the corresponding homopolymer **7**. Under special conditions, we obtained the AB type monomer **6**. However, after successful polycondensation it was impossible to transfer precursor **7** into polyelectrolyte **9**: due to the lack of solubilizing side chains in the activated intermediate **8**, this material proved to be completely insoluble and thus could not be converted into a constitutionally homogeneous product **9**. This failure forced a change in the synthetic strategy. We now combined the functions of the two different lateral substituents of **3**, *i.e.* solubilizing the polymer (done by the C_6H_{13} groups) and making possible final introduction of electrolyte functionalities (done by the CH_2-O-C_4H_9 groups), in one single type of side chains.[3] This was achieved by introducing long *n*-alkyl spacers between the PPP main chain and ether functionalities. Due to the longer spacers, however, the ether oxygen moved away from the benzylic position. To nevertheless maintain the required selectivity of the ether cleavage process – which must provide 100% halogenalkyl groups to prevent crosslinking – aliphatic-aromatic ethers were applied. Hexamethylene spacers proved to solubilize the rod-like macromolecules sufficiently even after ether cleavage, leading to the activated intermediate **11**. Hence, **10** was the precursor polymer of choice.

Quite surprisingly, however, all anionic polyelectrolytes prepared from **11** such as **12** and **13** proved to be insoluble in water or aqueous bases although their charge density was twice as high as in polymers like **5**.[3] In contrast to this, cationic polyelectrolytes such as **14 - 16**, easily available via conversion of **11** with a tertiary amine, were soluble not only in polar organic solvents but even in pure water.[4,5] We believe this is due to the fact that the apolar interior of these cylinder-like polyelectrolytes is covered by a sufficiently dense and homogeneous shell of hydrophilic cationic groups which prevent intermolecular hydrophobic interactions. PPP polyelectrolytes **15** and **16** could thus be analyzed with regard to their properties in solution using osmometry, small-angle X-ray scattering and viscosimetry. Special emphasis was directed towards the so-called "counterion condensation".

© 2004 WILEY-VCH Verlag GmbH & KGaA, Weinheim

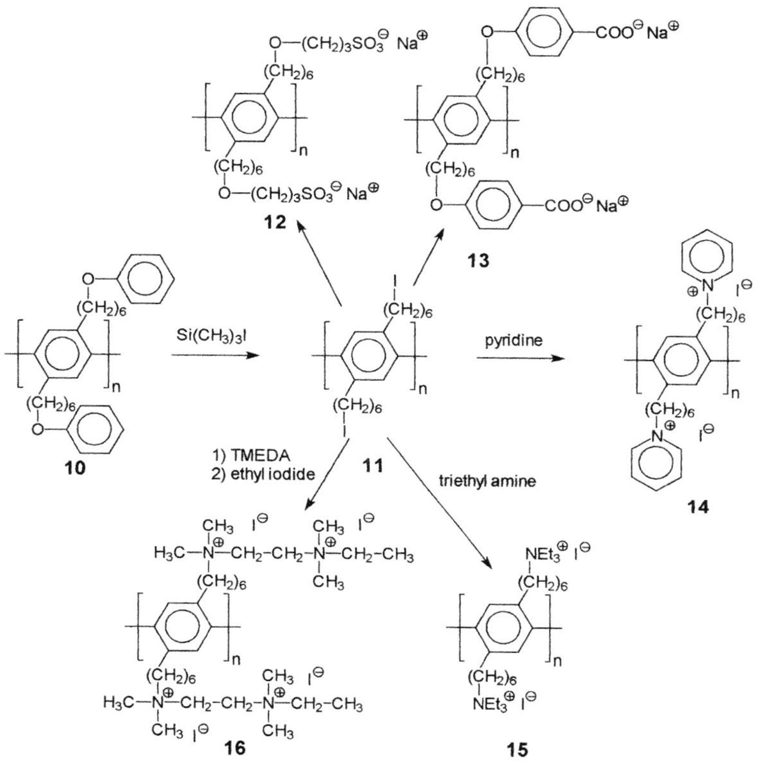

Scheme 3: Synthesis of PPP polyelectrolytes via 6-phenoxyhexyl-substituted precursor

Properties in Solution of PPPs 15 and 16

The phenomenon of counterion condensation was first described by Manning.[6,7] It appears because polyelectrolytes dissociate in polar solvents into highly charged macroions and a large number of oppositely charged small counterions. The high electric field of the macroions strongly couples to the counterions which thus partly neutralize the macroions. In turn, this strong correlation reduces the thermodynamic activity of the counterions. Only a certain fraction is osmotically active. Consequently, aqueous solutions of polyelectrolytes show a much lower osmotic pressure π than expected based on the number of counterions provided by the polyelectrolyte. This reduction in counterion activity can be expressed in terms of the osmotic

© 2004 WILEY-VCH Verlag GmbH & KGaA, Weinheim

coefficient Φ. This is the quotient of the experimentally observed osmotic pressure π_{obs} and the ideal osmotic pressure π_{id} calculated for a solution of counterions interacting neither with themselves nor with the macroion:

$$\Phi := \pi_{obs} / \pi_{id}$$

A large number of experimental studies of Φ is available in the literature, demonstrating that Φ is of the order of 0.2 to 0.3 for strongly charged polyelectrolytes and univalent counterions in the dilute regime.[8] A theoretical description of the obtained data can be carried out using the so-called Poisson-Boltzmann (PB) cell model: generation of a cell model is a common way of reducing the complicated many-body problem to an effective one-particle theory, *i.e.* the case of a single polyelectrolyte chain in a cell.[8-13] The basic idea is to partition the whole solution into small sub-volumes, each containing just one single macroion together with all its counterions. Since each sub-volume is electrically neutral, the electric field will *on average* vanish on the cell surface. By virtue of this construction, different sub-volumes are electrostatically decoupled to a first approximation. One may thus hope to factorize the partition function and reduce the problem to the treatment of one sub-volume called "cell". The shape of the cell is assumed to reflect the symmetry of the polyelectrolyte itself. For a solution of rod-like polyelectrolytes with density ρ_P and rod length L this is a cylindrical cell with the radius R being fixed by the condition $\pi R^2 L \rho_P = 1$ (Figure 1).

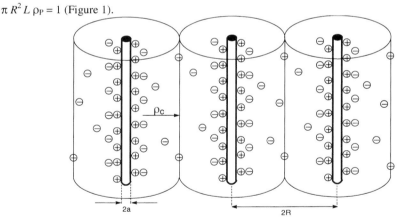

Figure 1:   Polyelectrolyte rods in solution as seen by the cell model

© 2004 WILEY-VCH Verlag GmbH & KGaA, Weinheim

As the theoretic treatment is again much simpler after neglecting end effects at the cylinder caps, the cylinder length is taken to be $L = \infty$ after mapping to the correct density. The ions are described as point-like, but they will electrostatically interact with the macroion as well as with each other. Their positions are thus strongly correlated. The solvent molecules are not explicitly taken into account: they are assumed to form a continuous dielectric background which is completely specified by its dielectric constant ε_r. In this simplified situation it can be proven that the osmotic pressure is defined by the counterion density at the cell boundary times the thermal energy $k_B T$.[14]

The obtained analytical description – taking into account all individual counterions – is still too involved since inter-ionic correlations complicate matters. The standard way out is to neglect these correlations in a mean-field spirit, as is done in the PB description: the ionic degrees of freedom are replaced by a cylindrical counterion density, and their interaction is approximated by the assumption that the density is locally proportional to the Boltzmann factor.[8-10,15] It can be shown that, on this mean-field level, the osmotic coefficient is defined by the ratio of boundary density and average density of the ions.[16]

For the PB treatment of the cell model, the rodlike polyelectrolyte main chain of radius a – having a contour distance per unit charge b – is coaxially enclosed in a cylinder of radius R. Electroneutrality is achieved by adding the appropriate amount of monovalent counterions, and no additional salt is present. The strength of the electrostatic interactions is conveniently expressed by the Bjerrum length

$$\lambda_B = e_0^2 / (4\pi\, \varepsilon_0\, \varepsilon_r\, k_B T)$$

where e_0 is the unit charge, ε_r is the dielectric constant of the solvent, and ε_0, k_B and T have their usual meanings.[8,17] This definition suggests a dimensionless way of measuring the line charge density of the rod via the charge parameter

$$\xi = \lambda_B / b.$$

ξ counts the number of unit charges on the rod per Bjerrum length and is usually called the Manning parameter. In the PB theory the osmotic coefficient is defined[16] by the expression

$$\Phi_{PB} = \frac{1+\gamma^2}{2\xi}$$

where the dimensionless constant of integration, γ, is the solution of the transcendental equation

© 2004 WILEY-VCH Verlag GmbH & KGaA, Weinheim

$$\gamma \ln \frac{R}{r_o} = \arctan \frac{1}{\gamma} + \arctan \frac{\xi - 1}{\gamma}.$$

In the limit of infinite dilution the cell radius R tends to infinity, which implies $\gamma \rightarrow 0$. For $\gamma \rightarrow 0$ and $\xi > 1$, the right-hand side of that equation tends to the constant π. Hence, the osmotic coefficient as computed by the PB theory (logarithmically) converges to the well-known Manning limiting law $\Phi_\infty = 1 / (2\xi)$. At finite densities it is always larger, however.

Unfortunately, this theoretical treatment suffers from various approximations. First, the cell model itself is a simplified representation of the polyelectrolyte solution: it neglects rod-rod interactions, is incapable of describing effects due to finite length of the rods, and reduces the solvent to a dielectric continuum. Second, the mean-field approach discards any inter-ionic correlations which can modify the average charge distribution. Furthermore, the rodlike macroions are confined in a periodic array of parallel, cylindrical Wigner-Seitz cells. On the other hand, this simplified representation greatly facilitates the solution of the PB equation. For salt-free solutions, the cell model provides an analytical expression for the distribution of the counterions around the macroion. Therefore we decided to compare the experimental results obtained by osmometry and small-angle X-ray scattering (SAXS) on the aqueous solutions of the PPP polyelectrolytes **15** and **16** with the predictions of the PB cell model.

For these investigations, it is of crucial importance that the PPP polyelectrolytes dissolve molecular-dispersly in water. Studies by SAXS (see below and Ref. [18]) and measurements of the electrical birefringence[19] demonstrate that this is the case at least in dilute solution. Moreover, the charge parameter ξ has to be known: its structural value is defined by $\xi = 3.4$ for polymer **15**, and $\xi = 6.8$ for polymer **16** (length of the repeating unit 0.43 nm, $\lambda_B = 0.73$ in water at 40 °C). The persistence length of the PPP main chain is high enough to regard these polymers as rodlike: the fully aromatic backbone of the PPPs exhibits a persistence length of approx. 20 nm.[20] The number-average contour length of PPP macroions used in the experiments is of the same order. Hence, the PPP polyelectrolytes may be treated as rod-like to a very good approximation.

© 2004 WILEY-VCH Verlag GmbH & KGaA, Weinheim

Osmotic measurements

Membrane osmometry is one of the methods that allow testing of the PB cell model. We consider the system of rodlike macroions **15**, dissolved in water without added salt. In this case the PB cell model can be solved exactly to yield the osmotic coefficient for finite concentrations. As input parameters, the cell model requires the charge parameter ξ and the length L of the macroions. The other parameter necessary for the calculation of Φ is the minimum distance of contact of the macroion and the counterions, termed a. This parameter is taken from SAXS analyses shown below which gave $a = 0.7$ nm.

The osmotic coefficients were measured for two kinds of counterions, iodide and chloride, in a membrane osmometer at 40 °C. The dilute solutions were prepared in water purified by reverse osmosis and ion exchange. As a representative example, Figure 2 shows the plot of the osmotic coefficient Φ *vs.* the polymer concentration for PPP **15**.

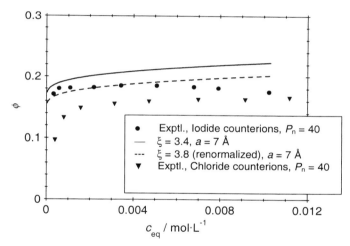

Figure 2:   Osmotic coefficient Φ as a function of counterion concentration c_{eq} for PPP **15**. The solid curve is the PB prediction of the cylindrical cell model, the dashed curve is due to the correlation-corrected PB theory

Both sets of experimental data clearly indicate that the PB solution of the cell model makes a fairly good prediction of the osmotic coefficient as the dominant change in Φ, a reduction of π by a factor of 5, is correctly accounted for: the PB theory predicts Φ to vary roughly within the range $0.18 - 0.22$, and the measured values accumulate around 0.18 (iodide) and 0.16 (chloride). Upon

© 2004 WILEY-VCH Verlag GmbH & KGaA, Weinheim

closer look, however, the PB theory is shown to systematically overestimate the osmotic coefficient: on the enlarged scale of Figure 2 it is visible that the measured values are systematically lower than the prediction, although still higher than the Manning limit $1 / (2\xi)$ of infinite dilution.

The fact that the PB theory overestimates the osmotic coefficient has been observed a number of times.[8,17] Careful studies of flexible polyelectrolytes indicated that agreement between the PB cell model and experimental data could only be achieved if the charge parameter ξ was renormalized to a higher value. The motivation given for this ad-hoc modification was the assumption of a local helical or wiggly main chain. Hence, the counterions "see" more charges per unit length, *i.e.* a macroion having a higher charge parameter. However, the results obtained for stiff chain macroions 15 show that the osmotic coefficient is lower than the PB results even for systems where the local conformation of the macroions is absolutely rodlike. Therefore, this common explanation cannot be true. A part of this discrepancy between theory and experiment seems to be due to the neglect of correlations in the mean-field approach, which will lower the osmotic pressure. Moreover, it became apparent that there are specific interactions of the counterions with the macroion which can be as large as the differences from the PB solution itself. This was concluded from the fact that chloride counterions lead to a considerably smaller Φ than iodide counterions, a result which cannot be explained on the basis of the PB cell model as it stands today.[8,17] This might be interpreted in terms of hydration, but cannot be taken into account explicitly in the framework of the PB cell model. The remaining discrepancy between theory and experiment should be looked at on the level of molecular details. This includes a better description of the solvent, hydration effects, or van der Waals forces.

SAXS investigations

The goal of the following SAXS investigation was to elucidate the spatial distribution of the counterions around the macroions, and to compare the results with the prediction of the PB cell model.[18] Small-angle X-ray scattering (SAXS) has been repeatedly used to study the counterion cloud around dissolved polyelectrolytes. In many cases the contribution of the macroion to the measured SAXS intensity $I(q)$ $\{q = [(4\pi/\lambda) \sin(\theta/2)]; \lambda$ = wavelength of the used irradiation; θ = scattering angle} exceeds by far the one of the counterions. The hydrocarbon backbone of polyelectrolytes **15** and **16**, however, exhibits a low excess electron density in water. Hence, their SAXS intensities are small, and the contribution of the counterions may become the leading term. By changing the counterions from iodide to chloride, moreover, the contrast can be changed quite drastically: while the electron density of chloride ions is nearly matched by water, iodide ions exhibit a strong contrast in aqueous solution. Hence, if iodide counterions are used, their correlation with the macroion should be easily visible. In the case of chloride counterions, on the other hand, the measured scattering intensity mainly originates from the macroions. Here, we focus our description on the iodide system.

The absolute scattering intensities $I(q)$ following from SAXS measurements may be rendered as

$$I(q) = \frac{N}{V} I_0(q) S(q)$$

where N/V is the number of dissolved polyelectrolyte molecules per volume. $S(q)$ is the structure factor which takes into account the effect of intermolecular interferences. Its influence is restricted to the region of small scattering angles. The scattering intensity $I_0(q)$ of a single rodlike polyelectrolyte molecule can be formulated as

$$I_0(q) = \int_0^1 [F(q,\alpha)]^2 d\alpha \qquad (1)$$

where α is the cosine of the angle between the scattering vector q and the long axis of the molecule. The scattering amplitude of the rod with orientation α follows as

$$F(q,\alpha) = L \frac{\sin(qaL/2)}{qaL/2} \int_0^\infty \Delta\rho(r_c) J_0[qr_c(1-\alpha^2)^{\frac{1}{2}}] 2\pi(r_c) dr_c. \qquad (2)$$

The measured scattering intensity $I_0(q)$ is thus related to the Hankel transform of the radial excess electron density $\Delta\rho(r_c)$ that can be calculated from the radial density of the macroions and the distribution of the counterions around the macroion.[18] Since the macroion has a small radius termed a, its radial electron density may be rendered in good approximation by a constant excess electron density $\Delta\rho_{rod}$. This quantity in turn can be derived from the partial specific volume of the macroion in solution.

The counterion distribution function $n(r_c)$ may be calculated from the solution of the PB equation within the frame of the cell model. Here, the polyelectrolyte is characterized by the charge parameter ξ. In order to obtain $n(r_c)$, the solution of uniform cylindrical polyelectrolytes of length

© 2004 WILEY-VCH Verlag GmbH & KGaA, Weinheim

L is treated as a system of N parallel rods (see Figure 1). The cell radius R follows from the number concentration N/V of the rods as $(N/V) \pi R^2 L = 1$. The distribution function $n(r_c)$ is defined by[21]

$$\frac{n(r_c)}{n(R_0)} = \left\{ \frac{2|\beta|}{\kappa(r_c)\cos[\beta \ln(r_c / R_M)]} \right\}^2 .$$

From the known parameters ξ, a and R, the first integration constant can be obtained through

$$\arctan\left(\frac{\xi-1}{\beta}\right) + \arctan\left(\frac{1}{\beta}\right) - \beta \ln\left(\frac{R_0}{a}\right) = 0 .$$

The second integration constant R_M may be regarded as the radial distance within which the counterions are condensed.[21] It follows that

$$R_M = a \exp\left[\frac{1}{\beta}\arctan\left(\frac{\xi-1}{\beta}\right)\right].$$

The screening constant κ and the number $n(R)$ of counterions at the cell boundary are related through $\kappa = 8 \pi \lambda_B n(R) = 4(1 + \beta^2) / R^2$. With $n(r_c)$ being known – $n(r_c)$ is the excess electron density within the cell, i.e. for $a \leq r_c \leq R$ – integration of equation (2) can be performed. The number of excess electrons per counterion $\Delta\rho_{ci}$ can be calculated to a good approximation by use of their crystallographic radii. $\Delta\rho(r_c)$ follows as $= \Delta\rho_{rod}$ for $0 \leq r_c \leq a$, as $= n(r_c)\cdot\Delta\rho_{ci}$ for $a < r_c \leq R$, and as $= 0$ for $r_c > R$. Insertion of these data in equation (2) followed by numerical integration of equations (2) and (1) then leads to the scattering intensity $I_0(q)$. The total scattering intensity of a system of non-interacting rods follows from the equation

$$I(q) = \frac{N}{V} I_0(q)S(q)$$

with $S(q) = 1$.

Figure 3 displays the absolute SAXS scattering intensities of PPP **15**.[18] In all cases, salt-free aqueous solutions with polymer concentrations c_P ranging from 3.52 up to 19.95 g·L^{-1} have been measured at 25 °C. The SAXS measurements are performed using a Kratky Kompakt camera. The scattering curves are normalized with regard to the polymer concentration (volume fraction of polymer Φ). The scattering intensities as a function of q obtained for all these solutions may be roughly divided into three regions. In the intermediate q-range, all curves decrease roughly as q^{-1}, whereas in the region of smallest values ($q < 0.4$ nm^{-1}) there is a strong rise of the measured

© 2004 WILEY-VCH Verlag GmbH & KGaA, Weinheim

scattering intensity $I(q)$. At highest q-values ($q > 2.8$ nm^{-1}), on the other hand, all curves deviate from the q^{-1} dependence and decrease more drastically.

For a fit according to the PB theory, most of the required parameters are fixed experimentally, including the charge parameter $\xi = 3.4$. The only adjustable parameter is the minimum distance to which the counterions may approach the macroion. In principle, this distance is the sum of the radii of counterion and macroion. In the cell model, however, the counterions are treated as point-like objects. Therefore, this minimum distance is defined just by the radius a of the macroion. In the following fit procedure, a is treated as an adjustable parameter. Here, data were taken only for the q-values greater than 0.49 nm^{-1}: trial calculations have shown that for $q > 0.49$ nm^{-1} the influence of mutual interactions of the polyelectrolytes may be safely dismissed.[18]

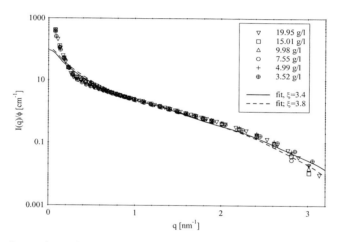

Figure 3: Comparison of measured and calculated SAXS scattering intensities for PPP **15**. The symbols give the experimental normalized scattering intensities of the aqueous solutions, the lines the optimal fits according to the PB cell model

Under these conditions, excellent agreement between theory and experiment can be achieved. Moreover, from the resulting fit, the radius a of the macroion follows as 0.7 nm. This value is considerably smaller than the value which would follow from the density of **15** in solution ($a \approx$ 1 nm). However, no meaningful description of the presented data is possible if $a = 1$ nm is assumed. This finding shows that the correlation of the counterion to the macroion is significantly

© 2004 WILEY-VCH Verlag GmbH & KGaA, Weinheim

stronger than predicted by the cell model: a smaller value of a obtained from the PB cell model calculations indicates a much higher concentration of the counterions directly at the surface of the macroion than anticipated from structural data of polyelectrolyte **15**. This is in accordance with the above measurements of the osmotic coefficient of **15** in aqueous solution, and here again, this underestimation of the macroion-counterion correlation might be taken into account by increasing the charge parameter to $\xi = 3.8$.

In the preceding discussion of the experimental data we assumed that the effect of mutual interactions between the macroions is restricted to the region of smallest scattering angles, *i.e.* $S(q) = 1$ was assumed in the region from where data were used for the fit according to the cell model. However, the effect of interaction seen in the experimental scattering curves at lowest q-values should be considered as well since it is evident that there is a strong deviation between theory and experiment for $q < 0.49$ nm^{-1} in Figure 3 which points to a long-range correlation of the macroions in solution, and indicates a weak attractive interaction between the rods. This effect is seen even better in the respective scattering curves of the fourfold charged PPP polyelectrolyte **16**:[22] Figure 4 shows an enlarged view of the data in the small-angle region of PPP **16**.

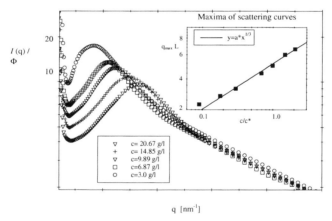

Figure 4:   SAXS intensities measured for PPP polyelectrolyte **16** at small scattering angles. The inset displays the position of the maximum as a function of the concentration

© 2004 WILEY-VCH Verlag GmbH & KGaA, Weinheim

A distinct maximum is evident similar to the "polyelectrolyte peak" found in salt-free solutions of flexible polyelectrolytes and of rod-like colloidal polyelectrolytes. Solutions of tobacco mosaic viruses, for example, show the position of the maximum q to depend sensitively on the concentration:[23] having defined the overlap concentration c^* of the rods through $c^* = L^{-3}$, q_{max} was found to be proportional to $(c/c^*)^{1/3}$ for concentrations smaller than c^*. For larger concentrations, the exponent $1/2$ was found. This was taken as evidence of the onset of partial aligning of the rods due to their increased mutual interaction. It is obvious from Figure 4 that the same phenomenon also occurs in salt-free solutions of the stiff-chain polyelectrolytes **16**: the maximum scales with $c^{1/3}$ in the dilute regime (see inset in Fig. 4) whereas an exponent of $1/2$ is found for higher concentrations. This suggests that there is some correlation in the orientation of the long axes of the polyelectrolyte molecules if the concentration exceeds c*.

Finally, at very small scattering angles ($q < 0.1$ nm^{-1}), there is a marked upturn of the scattering intensities. This could be shown to be a direct consequence of the respective upturn of $S(q)$. This may be assigned to weak attractive forces and thus correlation over the very long distances even in highly dilute systems. Hence, there is obviously some long-range interaction between the macromolecules even at very high dilutions.

Viscosity investigations

SAXS studies indicated some long-range interactions between the macromolecules even at very high dilutions. These interactions should be detectable in viscosity studies as well. Thus, viscosity measurements were carried out on salt-free and salt-containing (NEt$_4^+$I$^-$) aqueous solutions of PPP polyelectrolyte **15** using a capillary viscosimeter. The first question to be answered was whether or not the so-called "polyelectrolyte effect" will be observed in the Huggins plot. This effect is a matter of very controversial discussion in the literature.[24-28] There are papers which ascribe the whole effect to conformational changes, *i.e.* coil expansion as a consequence of increasing intramolecular coulomb repulsion at decreasing ionic strength. Several equations have been proposed to model the observed increase of η_{sp}/c_P at decreasing c_P such as those of Fuoss and Strauss.[29-31] However, the physical background of these equations is highly questionable because they do not describe the decrease of η_{sp}/c_P at very low c_P. Therefore, other concepts have been developed like that of Cohen, Priel and Rabin where the initial increase of η_{sp}/c_P at decreasing c_P

© 2004 WILEY-VCH Verlag GmbH & KGaA, Weinheim

followed by a decrease of η_{sp}/c_P at very low c_P is assigned to the intermolecular coulomb forces:[32-36] it is assumed that dilution of a polyelectrolyte solution starts at a rather high ionic strength. Thus, upon the first dilution steps, the Debye screening length increases faster than the mean distance of the individual macromolecules (Figure 5). As a consequence, the intermolecular electrostatic repulsion increases first and thus the reduced specific viscosity. Upon further dilution, the ionic strength is more and more determined by self-dissociation of water and dissolved gasses. Consequently, the screening length increases less efficiently now compared to the mean distance of the macromolecules. Consequently, the electrostatic interaction of the macroions decreases as well as the reduced intrinsic viscosity.

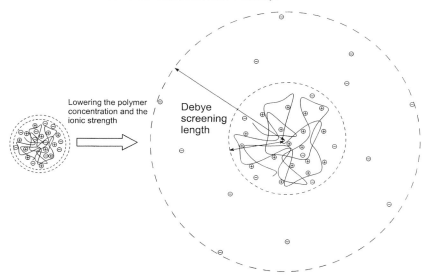

Figure 5:   Schematic picture of the development of the macroion's chain conformation (inner circle) and of the Debye screening length (outer circle) of flexible polyelectrolytes in salt-free aqueous solution upon dilution

In this latter theory the unusual behavior of the reduced specific viscosity is described as a consequence of electrostatics while in the former it follows from the chain's hydrodynamic volume. Therefore, it was important to show how the intrinsically rodlike macroions such as **15** behave. In order to visualize the results, the reduced intrinsic viscosities of PPP **15** obtained in aqueous solutions[37] were plotted *vs.* the polymer concentration (Figure 6). For the salt-free case,

© 2004 WILEY-VCH Verlag GmbH & KGaA, Weinheim

there is evidently a pronounced maximum of η_{sp}/c_P at $c_P \approx 5\times10^{-6}$ g·mL^{-1} while for increasing salt concentrations, this maximum becomes weaker and shifts towards higher polyelectrolyte concentrations. Finally, at salt concentrations as high as $c_s > 1\times10^{-3}$ g·mL^{-1} (not shown), the polyelectrolyte effect disappears completely, and a straight line results in the Huggins plot. Its extrapolation to $c_s = 0$ g·mL^{-1} gives an intrinsic viscosity of $[\eta] \approx 20$ mL·g^{-1}. This value is in excellent agreement with the intrinsic viscosity determined for the respective precursor PPP used for its preparation. Thus, at increasing salt concentrations, the viscosity behavior of PPP polyelectrolytes is in qualitative agreement with the behavior of flexible polyelectrolytes and naturally occurring rod-like polyelectrolytes such as DNA. However, there is a tremendous disagreement in the quantitative values, in particular at low salt concentrations.

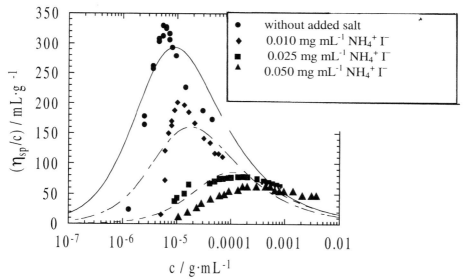

Figure 6:   Plot of the reduced specific viscosities, η_{sp}/c_P, vs. the concentration, c_P, of PPP polyelectrolyte **15**. The lines are the best fits of the experimental data according to the Cohen-Priel-Rabbin description and shown to make obvious the drastic deviation in particular for low ionic strengths

Also, it should be emphasized that it is obviously a general feature of rodlike PPP polyelectrolytes that their maxima in the Huggins plot appear at values of c_P approx. one order of magnitude smaller than those of flexible polyelectrolytes under the respective conditions. Beyond

© 2004 WILEY-VCH Verlag GmbH & KGaA, Weinheim

this, the maximum in the Huggins plot is much more pronounced and clearly narrower for PPP polyelectrolytes than for coiled macroions. The development of a detailed description of the viscosity behavior of PPP polyelectrolytes is a topic of future research. Nevertheless, it is currently clear that the so-called polyelectrolyte effect observed for flexible polyelectrolytes is certainly not the consequence of coil expansion but of long-range intermolecular electrostatic interactions.

Synthesis of PPP Polyelectrolytes via Amino Intermediates

All investigations carried out so far were performed using PPP polyelectrolytes having charge parameters ξ much larger than unity. Therefore counterion condensation was the predominant phenomenon in osmotic and scattering experiments. In order to develop a full understanding of polyelectrolytes, however, PPP polyelectrolytes with low charge parameters are required as well, *i.e.* having ξ equal to or even smaller than unity. Consequently, we prepared such PPP polyelectrolytes as well. For this purpose, the synthetic strategy shown in Scheme 4 was applied first.

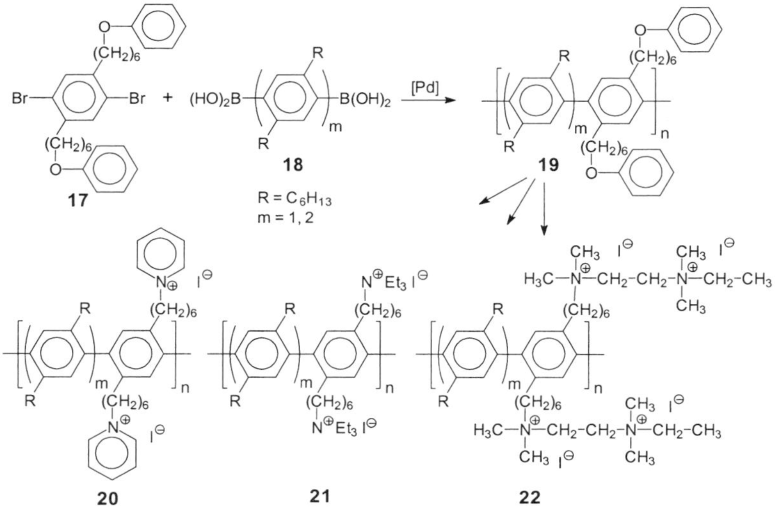

Scheme 4:   First syntheses of PPP polyelectrolytes having a lower charge parameter ξ

© 2004 WILEY-VCH Verlag GmbH & KGaA, Weinheim

Unfortunately, polyelectrolytes such as **20 – 22**, containing some phenylene moieties without charged side groups, proved to be insoluble in water. Insolubility in water, or dissolution only as well-defined aggregates, is also reported for similar PPP polyelectrolytes.[38-43] It seems that PPP polyelectrolytes in general need a very dense and homogeneous hydrophilic outer shell for molecular-disperse solubility in water. This shell must completely cover the cylinder-like polyions and thus suppress the hydrophobic interactions of the macromolecule's interior: if density and homogeneity of the hydrophilic shell are insufficient, intramolecular side-chain segregation occurs, leading to hydrophilic and hydrophobic areas on the macromolecule's outer shell. Experimental experience allows the conclusion that side-chain segregation and thus aggregation or precipitation of the polymers *via* hydrophobic interactions occur (*a*) when the apolar side chains are longer than the spacer groups between the polymer main chain and ionic side groups, (*b*) when the charged groups are only attached to one side of the phenylene moieties, or (*c*) in the case of polyelectrolytes such as **20 – 22** where only every second phenylene moiety bears charged side groups (Figure 7).

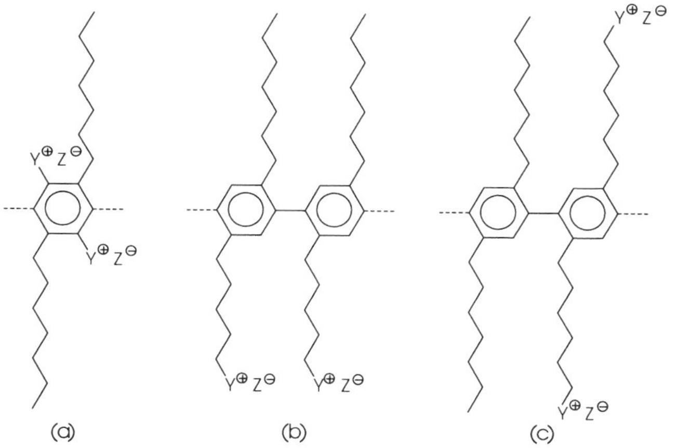

Figure 7:   General motifs leading to intermolecular hydrophobic interactions of rodlike PPP polyelectrolytes and thus insolubility in water

Water-soluble PPP polyelectrolytes with charge parameter $\xi \leq 1$ therefore need non-ionic building blocks that are hydrophilic enough to prevent aggregation. Thus we developed a new synthetic strategy for PPPs which should be soluble in water even at vanishing charge density.

© 2004 WILEY-VCH Verlag GmbH & KGaA, Weinheim

The key step here is the attachment of oligoethylene oxide (OEO) side chains to the PPP backbone.[44] These substituents should efficiently prevent hydrophobic interactions of the apolar PPP main chains. But the change from alkyl to OEO substituents as solubilizing side chains had some far-reaching consequences for the precursor strategy: OEO substituents do not allow the use of the above procedures where ether cleavage is a key step. Otherwise, this reaction would lead to the loss of the solubilizing side chains during precursor activation. The most convenient way to avoid ether cleavage as a macromolecular substitution step is to invert the original synthetic strategy, *i.e.* to use tertiary amines as the precursor functionalities, and to generate the polyelectrolytes via treatment of the precursor with low-molecular-weight alkyl halogenides or, alternatively, with acids (Scheme 5).[44]

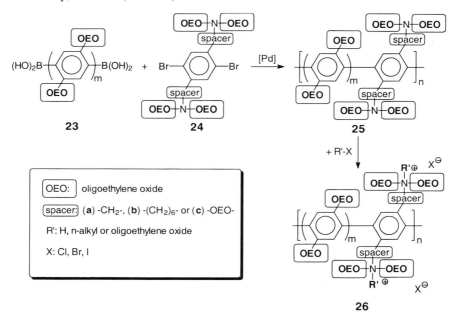

Scheme 5:   Improved synthesis of PPP polyelectrolytes having a lower charge parameter ξ

Using this new strategy, we obtained the first uncharged but nevertheless readily water-soluble precursor PPPs such as **27** ($P_n \approx 30$, see Scheme 6).

© 2004 WILEY-VCH Verlag GmbH & KGaA, Weinheim

Properties in Solution of Polybase 27

The new polyamine precursor PPP **27** represents a polybase because it contains amino groups in its side chains. Therefore it is of crucial importance to know the degree of protonation as a function of pH. Titration studies according to Scheme 6 show that this information is readily available using either ^{1}H NMR spectroscopy or potentiometry.

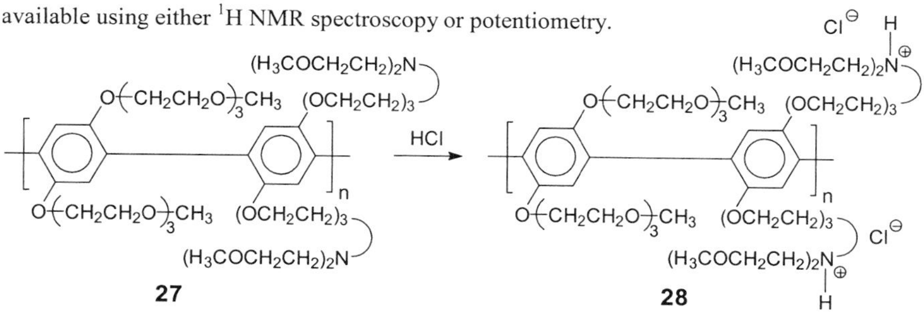

Scheme 6: Protonation of precursor PPP **27** using hydrochloric acid

It could be shown that protonation and deprotonation of PPPs **27** and **28**, respectively, take place in the range of approx. $10 > \text{pH} > 4$. The determined degrees of protonation are plotted against pH for precursor PPP **27** as well as for the tris(ω-methoxyethoxyethyl)amine which was used as a low-molecular-weight reference compound (Figure 8). The circles and dots are calculated from the experimental data determined using the pH electrode (values plotted at the left y-axis). The black triangles, on the other hand, are based upon the ^{1}H NMR chemical shifts of the respective N-CH$_2$ protons (δ-values plotted at the right y-axis). From these data, the buffer region was determined to be between $\text{pH} \approx 6.5$ and 8.5 for the monoamine, and between $\text{pH} \approx 6.0$ and 8.0 for polyamine **27**. Accordingly, the precursor PPP ($\text{pK}_a \approx 7.0$) seems to be a weaker base than the monoamine ($\text{pK}_a \approx 7.5$). Based on these results, we re-calculated the curves based on the Henderson-Hasselbalch's law of mass action for degrees of protonation between 10 % and 90 %. The results are represented in Figure 8 as the solid lines. In the case of the monoamine, the data determined using the pH electrode agree very well with the expected behavior of a monobasic amine having a pK$_a$ of 7.5. For the polymeric amine, on the other hand, there is obviously a slight deviation: the slope of the calculated curve is larger than that of the measured values. This is in agreement with previous experimental data obtained from flexible polybases. Also, it is in agreement with a model developed by Borkovec which allows the calculation of pK values of

© 2004 WILEY-VCH Verlag GmbH & KGaA, Weinheim

polybases on the basis of Ising models. Here, microscopic pK values are assumed for the individual acid-base groups of the molecule which depend on the (de)protonation state of their neighboring functional groups.[45,46]

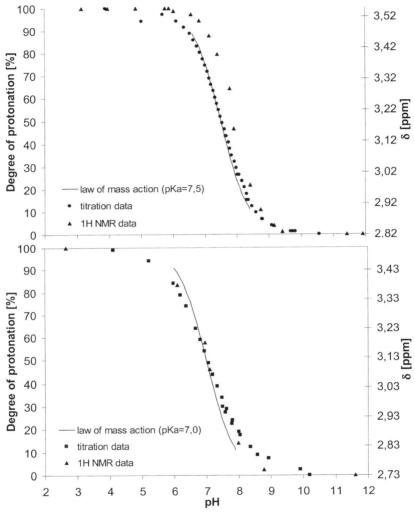

Figure 8:   Degrees of protonation and ^{1}H NMR chemical shifts of the monoamine (top plot) and of precursor PPP **27** (bottom plot) as a function of pH

© 2004 WILEY-VCH Verlag GmbH & KGaA, Weinheim

Conclusion

Efficient synthetic routes have been described for the preparation of rodlike PPP polyelectrolytes. Some of the obtained polymers dissolve in a molecular-disperse fashion even in pure water. For highly charged systems, the osmotic coefficient Φ proved to be a critical test for the PB cell model of rodlike polyelectrolytes. The measured data of Φ show a small but significant discrepancy compared to the PB theory. Similar conclusions were drawn from SAXS studies. These deviations may be traced back to the deficiency of the cell model which neglects ion-ion correlations as well as specific interactions between the macroions and its counterions. It became evident that the cell model nevertheless gives a semi-quantitative estimate of Φ – despite its simplicity. Moreover, for uncharged but nevertheless water-soluble PPP derivatives bearing amino groups in the side chains, it is found that the pK_a depends on the degree of protonation of the whole macromolecule. More detailed investigations concerning all these important phenomena are presently under way.

[1] I. U. Rau, M. Rehahn, *Macromol. Chem.* **1993**, *194*, 2225.
[2] I. U. Rau, M. Rehahn, *Polymer* **1993**, *34*, 2889.
[3] I. U. Rau, M. Rehahn, *Acta Polymerica* **1994**, *45*, 3.
[4] G. Brodowski, A. Horvath, M. Ballauff, M. Rehahn, *Macromolecules* **1996**, *29*, 6962.
[5] M. Wittemann, M. Rehahn, *J. Chem. Soc., Chem. Commun.* **1998**, 623.
[6] G. S. Manning, *J. Chem. Phys.* **1969**, *51*, 924, 934, 3249.
[7] G. S. Manning, *Ann. Rev. Phys. Chem.* **1972**, *23*, 117.
[8] M. Deserno, C. Holm, J. Blaul, M. Ballauff, M. Rehahn, *Eur. Phys. J.* **2001**, *E 5*, 97 and given references.
[9] R. M. Fuoss, A. Katchalsky, S. Lifson, *Proc. Natl. Acad. Sci. USA* **1951**, *37*, 579.
[10] T. Alfrey, P. W. Berg, H. Morawetz, *J. Polym. Sci.* **1951**, *7*, 543.
[11] A. Katchalsky, *Pure Appl. Chem.* **1971**, *26*, 327.
[12] K. S. Schmitz, *Macroions in solution and in colloidal suspension*, VCH Publishers, New York, 1993.
[13] M. Mandel, *J. Phys. Chem.* **1992**, *96*, 3934.
[14] L. Wennerström, B. Jönsson, P. Linse, *J. Chem. Phys.* **1982**, *76*, 4665.
[15] M. Deserno, C. Holm, S. May, *Macromolecules* **2000**, *33*, 199.
[16] R. A. Marcus, *J. Chem. Phys.* **1955**, *23*, 1057.
[17] J. Blaul, M. Wittemann, M. Ballauff, M. Rehahn, *J. Phys. Chem B* **2000**, *104*, 7077 and given references.
[18] B. Guilleaume, J. Blaul, M. Wittemann, M. Rehahn, M. Ballauff, *J. Phys.: Condens. Matter* **2000**, *12*, A245 and given references.
[19] K. Lachenmeyer, W. Oppermann, manuscript in preparation.
[20] P. Galda, *Dissertation*, Karlsruhe, 1994.
[21] M. LeBret, B. Zimm, *Biopolymer* **1984**, *23*, 287.
[22] B. Guilleaume, J. Blaul, M. Ballauff, M. Wittemann, M. Rehahn, G. Goerigk, *Eur. Phys. J.* **2002**, *E 8*, 229 and given references.
[23] E. E. Maier, R. Krause, M. Deggelmann, M. M. Hagenbüchle, R. Weber, S. Fraden, *Macromolecules* **1992**, *25*, 1125.
[24] S. Förster, M. Schmidt, *Adv. Polym. Sci.* **1995**, *120*, 53.
[25] W. Oppermann, *Makromol. Chem.* **1988**, *189*, 927.
[26] G. Weill, *J. Phys. (France)* **1989**, *49*, 1049.

© 2004 WILEY-VCH Verlag GmbH & KGaA, Weinheim

[27] S. Förster, M. Schmidt, M. Antonietti, *Polymer* **1990**, *31*, 781.
[28] Y. Yamanaha, H. Matsuoka, M. Hasegawa, N. Ise, *J. Am. Chem. Soc.* **1990**, *112*, 587.
[29] R.M. Fuoss, U.P. Strauss, *J. Polym. Sci.* **1948**, *3*, 246.
[30] R.M. Fuoss, U.P. Strauss, *J. Polym. Sci.* **1948**, *3*, 603.
[31] R.M. Fuoss, U.P. Strauss, *J. Polym. Sci.* **1949**, *4*, 96.
[32] J. Cohen, Z. Priel Y. Rabin, *J. Chem. Phys.* **1988**, *88*, 7111.
[33] J. Cohen, Z. Priel, *Macromolecules* **1989**, *22*, 2356.
[34] J. Cohen, Z. Priel, *Polym. Commun.* **1989**, *30*, 223.
[35] J. Cohen, Z. Priel, *J. Chem. Phys.* **1990**, *93*, 9062.
[36] W. Hess, R. Klein, *Adv. Phys.* **1983**, *32*, 173.
[37] M. Wittemann, *Dissertation*, Karlsruhe, 1999.
[38] R. Rulkens, M. Schulze, G. Wegner, *Macromol. Rapid Commun.* **1994**, *15*, 669.
[39] S. Vanhee, R. Rulkens, U. Lehmann, C. Rosenauer, M. Schulze, W. Köhler, G. Wegner, *Macromolecules* **1996**, *29*, 5136.
[40] R. Rulkens, G. Wegner, T. Thurn-Albrecht, *Langmuir* **1999**, *15*, 4022.
[41] P. Baum, W. H. Meyer, G. Wegner, *Polymer* **2000**, *41*, 965.
[42] M. Bockstaller, W. Köhler, G. Wegner, D. Vlassopoulos, G. Fytas, *Macromolecules* **2000**, *33*, 3951.
[43] M. Bockstaller, W. Köhler, G. Wegner, D. Vlassopoulos, G. Fytas, *Macromolecules* **2001**, *34*, 6359.
[44] S. Traser, P. Wittmeyer, M. Rehahn, *e-Polymers,* **2002**, no. 032.
[45] M. Borkovec, G, J. M. Koper, *J. Phys. Chem.* **1994**, *98*, 6038.
[46] G. J. M. Koper, M. Borkovec, *J. Phys. Chem. B* **2001**, *105*, 6666.

© 2004 WILEY-VCH Verlag GmbH & KGaA, Weinheim

Particle Scattering Factor of Pearl Necklace Chains

Ralf Schweins,[1,2] *Klaus Huber*[*][1]

[1]Universität Paderborn, Fakultät für Naturwissenschaften, Department Chemie, Warburger Str.100, D-33098 Paderborn, Germany
E-mail: huber@chemie.uni-paderborn.de
[2]Institut Laue - Langevin, Large Scale Structures Group, B. P. 156, 6, rue Jules Horowitz, F-38042 Grenoble CEDEX 9, France

Summary: The particle scattering behaviour of a pearl necklace chain is derived. The chain is composed of sphere-like pearls, separated by rod-like segments of fixed length, which have no angular restrictions. By calculating several series of model scattering curves, the important structural features are retrieved. The model is believed to be useful in interpreting intermediate structures of collapsing macromolecules or polyelectrolytes. A first application to a shrinking polyelectrolyte coil generated by molecular dynamic simulations (Limbach and Holm, J.Phys.Chem. 2003) is presented and used to discuss the potentials and limits of the model.

Keywords: coil collapse; form factor; freely jointed chain; pearl necklace; polyelectrolyte collapse

Introduction

Polyelectrolytes are highly charged and water soluble macromolecules. Discharging the polyelectrolytes reduces their solubility. If water is a non-solvent to the respective neutral polymer backbone, electrical discharge eventually leads to a precipitation of a polyelectrolyte salt. This neutralisation of the chains can be caused by protonation, by complexation of metal cations[1-3] or by addition of a large amount of an inert salt.[3-5] Very low concentrations of the polyelectrolyte chains may prevent a macroscopic phase separation for kinetic reasons resulting in an intramolecular collapse to sphere like particles. Predicted by various theoretical approaches, this collapse was established experimentally for the first time with sodium polyacrylate chains (NaPA) in the presence of calcium ions.[6,7]

© 2004 WILEY-VCH Verlag GmbH & KGaA, Weinheim   DOI: 10.1002/masy.200450702

A similar coil collapse occurs with neutral polymers when crossing the unperturbed or Θ-state.[8-13] First time resolved experiments by means of dynamic light scattering on the collapse mechanism of neutral polymers revealed a two stage kinetics with a crumpled globule in a first stage and a final collapse to a compact globule or sphere. The collapse was inferred by quenching polymer solutions from the Θ-temperature to just below its separation threshold. The time regime for the induced collapse was in the order of a few minutes and required extremely skillful experiments.[12]

Much less experimental data exist on the shrinking mechanism of polyelectrolyte coils close to a phase boundary. In the latter case, however, theoretical progress is remarkable. Consideration of the polyelectrolyte collapse began with the publications by Khokhlov,[14] Kantor and Kardar[15,16] and Rubinstein et al.[17] The underlying physics is the shape instability of charged droplets which depends on the surface tension and charge density. However, unlike to charged droplets, a collapsed chain cannot fall apart. Its conformational transformation from spherically collapsed particles to extended polyelectrolyte coils or vice versa passes a cascade of transition states. Depending on the conditions, these states may adopt cigar like or pearl necklace like structures.[15-22] The transformation can be induced by addition of an increasing amount of an inert salt to a polyelectrolyte, dissolved in a medium which is a bad solvent for its neutral backbone.

First experimental indications for pearl necklace like intermediates adopted by polyelectrolyte chains in dilute solution were published only recently. Geissler et al.[23] induced the shrinking of a polycation in saltless water by addition of acetone and performed small angle neutron scattering (SANS) experiments with collapsed chains. By comparing the overall size of the polycations with the behavior of the scattering curves at high q-values, they found indication for a string of three to four pearls. Similar to this type of shrinking, Morawetz et al.[24] followed solutions of sodium polystyrene-sulfonate, sodium polyacrylate (NaPA) and sodium polymethacrylate (NaPMA) close to the phase boundary by adding methanol to the aqueous solutions and performed NMR spectroscopy. They found a reduction of the ^{1}H signals which was attributed to the loss of mobile segments. According to Morawetz et al.[24], those segments were used up to form the pearls. The most recent paper in this field[25] also seems to provide the most direct indications. Minko et al. succeeded to produce AFM images which indicate a cascade of structural transitions of poly(2-vinylpyridine) in solution, induced by Pd^{2+} complexation. The cascade started with wormlike chains and finally led to pearl necklace

© 2004 WILEY-VCH Verlag GmbH & KGaA, Weinheim

like structures. At the same time first systematic small angle scattering has been carried out on highly dilute solutions of NaPA in the presence of earth alkaline cations.[26,27] These make the development of theoretical scattering curves for pearl necklace like structures highly desirable.

First theoretical scattering curves[20-22] stem from computer simulations of the collapse process of polyelectrolyte chains induced by counterion condensation. A characteristic feature of those curves is a shoulder or a maximum of the scattering curve at a distinct value of the scattering vector. This value was related to the distance between neighbored pearls.

Initial attempts to calculate analytical scattering curves have been made by Rubinstein et al.[17] and by Francois et al.[28] In Ref. 17, pearls were lined up along a rodlike string. In Ref. 28, the pearls were assumed to be connected by Gaussian chain segments which are modeled by bonds with a Gaussian length distribution without angular correlations. The two models limit the range of flexibility reaching from a rod like arrangement of pearls to a flexible chain of pearls. In both cases, the scattering from strings was neglected which may be justified in the light of the much larger mass fractions being located in the pearls. However, it may become an increasingly crude approximation if the interconnecting strings are coiled segments.

We therefore extend these calculations by a model which is characterized by features. (1) We connect rods of constant length denoted as strings to form a freely jointed chain with pearls on its junctions. (2) We explicitly include the scattering contribution from the interconnecting strings. An appropriate assessment of both features seems to be desirable and is presented in the following. Aside from the explicit consideration of the interconnecting strings, the present model differs from the suggestion by Rubinstein et al.[17] in its angular flexibility. However, we have to emphasize, that the adoption of a certain extent of angular flexibility seems to be realistic only as long as two next but one pearls do not get closer than the length of a string in accord with the Rayleigh instability. A certain extent of angular flexibility may also become a realistic feature for pearl necklace structures which have causes[29] different from the Rayleigh instability of polyelectrolytes. At the same time, our model fixes the string length. This may be a more realistic feature than the distribution width of an interconnecting Gaussian coil segment in the model of Francois et al.,[28] at least if the separation of charged pearls along a polyelectrolyte chain is addressed. Thus, the present model accounts for aspects not

© 2004 WILEY-VCH Verlag GmbH & KGaA, Weinheim

considered by the previous calculations[17,28] without being necessarily superior in all respects. However, it is the application of the different models which may help to distinguish the consequences of various features on the scattering of intermediates of a collapsing polymer.

In the first part of the paper, we describe the model used for the pearl necklace and derive its particle scattering factor. In the second part, an outline is given for the major aspects of the scattering behavior by discussing the impact of systematically varied parameters. Finally, the model is compared with well defined structures Limbach and Holm[22] have generated by means of molecular dynamics simulations.

Calculation of Particle Scattering Factor

The Model. The pearl necklace chain is composed of two elements: Rod like strings are interconnected at their ends to form a freely jointed chain. Each linking point is at the center of a homogeneous sphere corresponding to a pearl. M rod like strings connect $M + 1 = N$ pearls. The overall contour length of the chain is

$$L = M \cdot A \tag{1}$$

with A the fixed distance between centers of two neighboring pearls. No angular correlations exist between any two connected rod like strings. The radius of the pearls is R which reduces the physical length of the strings to l

$$l = A - 2R \tag{2}$$

A detailed illustration of the model is given in Figure 1.

The model can be mapped to any chemically specified macromolecular chain according to

$$m_r = l / b \tag{3a}$$

$$m_s = 4\pi R^3 / 3v_o \tag{3b}$$

© 2004 WILEY-VCH Verlag GmbH & KGaA, Weinheim

which relates the number of monomers in a rod like segment m_r to the length b of the chemical monomer and the number of monomers in a pearl m_s to the monomer volume v_o respectively. The total mass M_w of the polymer chain thus amounts to

$$M_w = M \cdot m_r + N \cdot m_s \qquad (4)$$

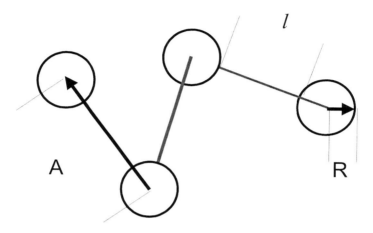

Figure 1. A pearl necklace based on a freely jointed chain with A the distance between the centers of two neighboring pearls, l the nearest distance between the surfaces of two neighboring pearls and R the a radius of a pearl.

Elementary Scattering Functions. Independent of the nature of the applied waves, the scattering pattern exerted by a scattering particle can be expressed by the form factor $P(q)$

$$P(q) = I(q)/I(q = 0) \qquad (5)$$

The form factor corresponds to the scattering intensity $I(q)$ as a function of the scattering vector q, normalized by the scattering intensity in forward direction $I(q = 0)$. The scattering vector q is defined as

$$q = (4\pi/\lambda)\sin(\theta/2) \qquad (6)$$

© 2004 WILEY-VCH Verlag GmbH & KGaA, Weinheim

with θ the scattering angle and λ the wavelength of the scattered waves.

A suitable starting point for the calculation of particle scattering factors was published in 1915, when Debye first predicted the scattering behavior of small molecules.[30] At that time corresponding scattering curves were about to become accessible by X-ray diffraction. Treating molecules as clusters of atoms with fixed inter-atomic distances r_{ij}, Debye expressed the form factor as a double sum over all atomic pair combinations

$$P(q) = \frac{1}{n^2} \sum_{i=1}^{n} \sum_{j=1}^{n} \psi_i(q)\psi_j(q) \frac{\sin(qr_{ij})}{qr_{ij}} \tag{7}$$

In Equation (7) the factor $\psi_i(q)$ is the normalized scattering amplitude of the atom i.

In 1969 this formula was used to calculate the form factor of n spheres, lined up on a freely jointed chain. Although, Burchard's and Kajiwara's[31] original intention was to describe the scattering behavior of macromolecules, their results turn out to be an important element for the calculation of the scattering behavior of a pearl necklace. In their model, the spheres corresponded to monomeric units which were connected via bonds of constant length. The bonds fixed the distance between neighboring monomers and formed the freely jointed chain. All monomers were considered to be identical and their scattering behavior could be described by a single normalized scattering amplitude $\psi_i(q) = \psi(q)$. Due to the fact that a sphere is invariant to orientations, the form factor of a sphere is equal to the product $\psi(q)\psi(q)$. Scattering contribution from the bonds were neglected. The resulting scattering curve corresponded to a modified Equation (7)

$$P(q) = \frac{1}{n^2} \psi^2(q) \sum_{i=1}^{n} \sum_{j=1}^{n} \left\langle \frac{\sin(qr_{ij})}{qr_{ij}} \right\rangle \tag{8}$$

In Equation (8) n is the degree of polymerization of the chain. The brackets $< >$ denote an averaging of the interference factor from two point scatterers, j and i, over all possible conformations of the connecting chain segment with |j-i| monomers. For freely jointed chain segments, this conformational average reads[33]

© 2004 WILEY-VCH Verlag GmbH & KGaA, Weinheim

$$\left\langle \frac{\sin(qr_{ij})}{qr_{ij}} \right\rangle = \left(\frac{\sin qb}{qb} \right)^{|j-i|} \tag{9}$$

In Equation (9) b is the distance between neighboring monomers, corresponding to the bond. By inserting Equation(9) into Equation (8) and solving the double sum, Burchard and Kajiwara reached at

$$P(q) = \frac{2\psi^2(q)}{n^2} \left[\frac{n}{1 - \sin(qb)/qb} - \frac{n}{2} - \frac{1 - (\sin(qb)/qb)^n}{(1 - \sin(qb)/qb)^2} \cdot \frac{\sin(qb)}{qb} \right] \tag{10}$$

A second basic element required for the form factor of a pearl necklace can be recovered from a paper by Hermans and Hermans.[32] Like Burchard and Kajiwara,[31] Hermans' and Hermans' intention was to model the scattering curve of polymeric chains. However, in their case, the monomers were the infinitely thin rods or bonds which form the freely jointed chain. Due to the fact that no angular correlation does exist between two neighboring monomeric bonds, the orientational averaging of each bond was performed independently of the rest of the chain. Starting point is again Equation (8). The scattering behavior of the monomers was expressed by the normalized scattering amplitude of an infinitely thin rod as $\psi_i(q) = \Lambda(q)$. The amplitude $\Lambda(q)$ is an orientational average. Hermans and Hermans could also apply Equation (9) to account for the averaging over all possible distances between monomer i and j. In line with the nomenclature of Equation (8) to Equation (10), we use a chain with m = n-1 bonds each of constant length b. It has to be emphasized that, unlike for spheres, the form factor of a rod is not equal to $\Lambda^2(q)$. The final form of the particle scattering function was[32]

$$P(q) = \frac{\Lambda^2(q)}{m^2} \left[m \left\{ 2/\Lambda(q) - \left(\frac{\sin(qb/2)}{\Lambda(q)qb/2} \right)^2 \right\} + \frac{2m}{1 - \sin(qb)/qb} - 2\frac{1 - (\sin(qb)/qb)^m}{(1 - \sin(qb)/qb)^2} \right] \tag{11}$$

Form Factor of a Pearl Necklace Chain. The particle scattering function of a pearl necklace presented in this paper is based on the freely jointed chain which determines the distances between any two pearls, each located on the ankle of two neighboring bonds (strings). The

© 2004 WILEY-VCH Verlag GmbH & KGaA, Weinheim

scattering curve P(q) decomposes into three contributions. Two of these contributions can be adopted from the elements presented in the preceding section. The term $S_{ss}(q)$ corresponds to the interferences stemming from the pearls and $S_{rr}(q)$ quantifies the interferences of the interconnecting rod like strings. The third contribution $S_{rs}(q)$ is a mixed term, accounting for correlations between the interconnecting strings and the pearls which, to the best of our knowledge, is evaluated for the first time in the present work. Addition of all three contributions and appropriate normalization leads to the form factor of a pearl necklace P(q)

$$P(q) = \frac{S_{ss}(q) + S_{rr}(q) + S_{rs}(q)}{(M \cdot m_r + N \cdot m_s)^2} \tag{12}$$

Normalization is achieved by the total number of chemical monomeric units forming the polymer chain.

The term $S_{ss}(q)$ can be translated from Equation (10) by simply renaming b to A and n to N and by weighting with the amount of monomers $(m_s N)^2$.

$$S_{ss}(q) = 2m_s^2 \psi^2(q) \left[\frac{N}{1 - \sin(qA)/qA} - \frac{N}{2} - \frac{1 - (\sin(qA)/qA)^N}{(1 - \sin(qA)/qA)^2} \cdot \frac{\sin(qA)}{qA} \right] \tag{13}$$

The normalized scattering amplitude for a pearl will be based on a sphere according to[34]

$$\psi(q) = \left[3 \cdot \frac{\sin(qR) - (qR) \cdot \cos(qR)}{(qR)^3} \right] \tag{14}$$

Using the correspondence of m and M, Equation (11) leads to the contribution from the rod like segments to the scattering of a pearl necklace. Here the weighting factor is based on the amount of monomers in all rods $(m_r M)^2$

$$S_{rr}(q) = m_r^2 \left[M \left\{ 2\Lambda(q) - \left(\frac{\sin(ql/2)}{ql/2} \right)^2 \right\} + \frac{2M\beta^2(q)}{1 - \sin(qA)/qA} - 2\beta^2(q) \frac{1 - (\sin(qA)/qA)^M}{(1 - \sin(qA)/qA)^2} \right] \tag{15}$$

© 2004 WILEY-VCH Verlag GmbH & KGaA, Weinheim

However, we have to draw attention to an important difference to the Hermans and Hermans model. In the latter case, the scattering of the rod like segments stems from rods with the full length A which is the distance between two neighboring ankles. In the pearl necklace, any connecting rod ends at the surface of two neighboring pearls reducing the effective length to $l = A-2R$. The remaining parts of A were attributed to the pearls. Thus, two different expressions are required in Equation (15) in order to account for the scattering amplitudes of the rod like segments. The curved bracket is the self term, representing the form factor of an infinitely thin rod of length l. Here $\Lambda(q)$ denotes the orientationally averaged and normalized scattering amplitude of the rod like segments given as[35]

$$\Lambda(q) = \frac{\int\limits_{0}^{ql} \frac{\sin(t)}{t}dt}{ql} \tag{16}$$

In the remaining two terms of Equation (15), the expression $\beta(q)$ is used to form products of amplitudes corresponding to pair combinations of any two rods which start at R and end at A-R if described in the coordinate system of the preceding pearl respectively

$$\beta(q) = \frac{\int\limits_{qR}^{q(A-R)} \frac{\sin(t)}{t}dt}{ql} \tag{17}$$

Finally, the mixed term remains to be evaluated. This term includes only a part of the double sum in Equation (7). In this part, the index i and j denotes pearls and rod like strings respectively. The scattering amplitude $\psi_i(q) = \psi(q)$ of pearl i is given by Equation (14) and the orientationally averaged and normalized scattering amplitude $\psi_j(q) = \beta(q)$ of a rod j is defined by Equation (17). Casting this term into a double sum and using Equation (9) with A instead of b leads to the mixed term

$$S_{rs}(q) = m_r \beta(q) \cdot m_s \psi(q) \cdot 2 \left[\sum_{i}^{N} \sum_{j}^{M} \left(\frac{\sin(qA)}{qA} \right)^{\alpha_{ij}} \right] \tag{18}$$

Care has to be taken in properly expressing the exponent α_{ij} which reads

© 2004 WILEY-VCH Verlag GmbH & KGaA, Weinheim

$$\alpha_{ij} = |i\text{-}j| \qquad j \geq i \qquad\qquad (19a)$$

$$\alpha_{ij} = |i\text{-}j\text{-}1| \qquad j < i \qquad\qquad (19a)$$

It can easily be shown, that the number of pearl-rod combinations separated by k intermediate rods amounts to 2(N-1-k). Thus by using Equation (19), Equation (18) can be rewritten by the following sum

$$S_{rs}(q) = m_r \beta(q) \cdot m_s \psi(q) \cdot 2 \left[2\sum_{k=0}^{N-2} (N-1-k)\left(\frac{\sin(qA)}{qA}\right)^k \right] \qquad (20)$$

The sum in Equation (20) can be solved, leading to

$$S_{rs}(q) = m_r \beta(q) \cdot m_s \psi(q) \cdot 4 \left[\frac{N-1}{1-\sin(qA)/qA} - \frac{1-(\sin(qA)/qA)^{N-1}}{(1-\sin(qA)/qA)^2} \cdot \frac{\sin(qA)}{qA} \right] \quad (21)$$

Inserting Equation (13), (15) and (21) into Equation (12) finally leads to the form factor of a pearl necklace based on a freely jointed chain.

Parameters. Sodium polyacrylate was chosen as an example for a polyelectrolyte system which is able to collapse to spheres via non-spherical intermediate states.[7,26] A pearl necklace model can be adapted to this system by using a monomer length of b = 2.591 Å and a monomeric volume of v_o = 586.2 Å^3. The latter was estimated from the hydrodynamically effective radius R_h = 170 Å of a completely collapsed sodium polyacrylate chain with a molar mass of $3.3\,10^6$ Dalton[26,36].

Results

In order to estimate the impact of different structural elements of a pearl necklace on its particle scattering factor, three series of scattering curves were generated.

© 2004 WILEY-VCH Verlag GmbH & KGaA, Weinheim

The first series addresses to the influence of the distance between neighboring pearls at a constant pearl size of R = 80 Å. The distance was varied between 335 Å < A < 510 Å. Results are summarized in Figure 2. All curves exhibit a minimum. The scattering vector of the minimum q_{min} decreases with increasing distance between the neighboring pearls A. In order to estimate a pearl distance A, use can be made from the following empirical relationship

$$A = 10.67 \cdot q_{min}^{-0.83} \tag{22}$$

A second minimum does not become noticeable due to the lack of correlation between pearls separated by two strings. Above q = 0.03 Å^{-1}, all three curves merge and the scattering is governed solely by the individual pearls.

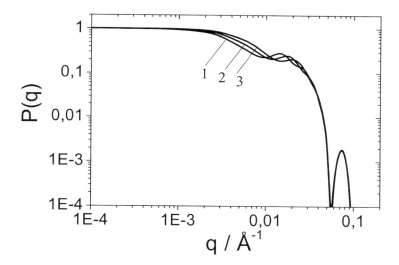

Figure 2. Form factors of pearl necklace chains with three pearls. The pearl size is fixed at R = 80 Å. The distance between two pearls is A = 510 Å (1); A = 410 Å (2) and A = 335 Å (3).

As shown in Figure 3, Equation (22) is reliable in a regime of A/R which is at least as large as 3 < A/R < 6 .

© 2004 WILEY-VCH Verlag GmbH & KGaA, Weinheim

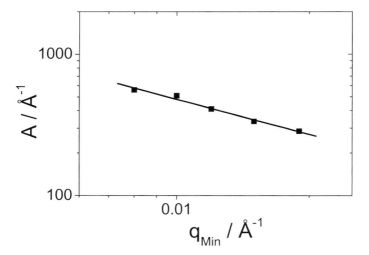

Figure 3. The distance A between the center of two neighboring pearls plotted versus the location of the minimum observed for model scattering curves of trimpbells. The pearl size was fixed at R = 80 Å. The straight line is a fit with Equation (22).

In a second series, the pearl distance is fixed at A = 500 Å and scattering curves are calculated for a variable pearl size in a range of 75 Å < R < 200 Å. Now, the location of the minimum q_{min} remains unchanged at least within this series, but the minimum is getting blurred with increasing R/A and eventually turns into a shoulder. In the case of a shoulder, A could only be estimated by Equation (22) if the inflection point is used instead of q_{min}. However, as shown in Figure 4, another distinct feature becomes discernible. The sharp minimum at q_s moves towards higher q-values for a decreasing pearl size. This could be captured by the empirical equation,

$$R = 4.4 / q_s \qquad (23)$$

which is almost identical to the well known relation $R = 3\pi/2q_s$ for spheres.[34] Clearly, q_s can be attributed to the sphere radius R. Alternatively, the respective maxima to the right of q_s could have equally well been chosen to determine R.

Finally, a series of scattering curves was calculated for variable scattering power of the pearl and rod. This variation was achieved by changing b and v_o, which, according to Equation(3),

© 2004 WILEY-VCH Verlag GmbH & KGaA, Weinheim

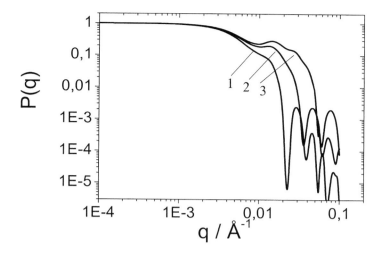

Figure 4. Form factors of pearl necklace chains with three pearls. The distance between neighboring pearls is fixed at A = 500 Å. The pearl size is R = 200 Å (1); R = 125 Å (2) and R = 75 Å (3).

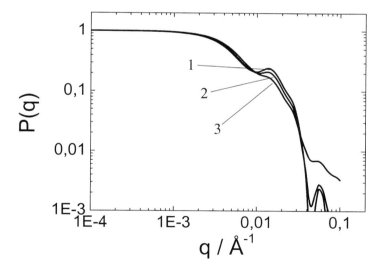

Figure 5. Form factors of pearl necklace chains with three pearls at variable b and v_o. The pearl size and the distance between two neighboring pearls is fixed at R = 100 Å and A = 500 Å respectively. The variation of b and v_o changes the ratio of monomers in a pearl m_s to the number of monomers in an interconnecting rod m_r according to m_s / m_r = 7146/116 (1); m_s / m_r = 6500/1083 (2) and m_s / m_r = 5416/2708 (3).

© 2004 WILEY-VCH Verlag GmbH & KGaA, Weinheim

simply varies the number of monomers per rod m_r and per pearl m_s. In order to isolate this aspect, calculation of the series was performed at a constant distance A between two neighboring pearls and at constant pearl size R. Results are represented in Figure 5. In agreement with the earlier series, the location of the first minimum q_{min} is not affected by the present changes. But the first minimum, which is related to A is smeared out as the ratio m_r/m_s increases and so does q_s.

To conclude, the most prominent feature of our model scattering curves is the minimum q_{min} referring to the distance A between two neighboring pearls. Although q_{min} may not be related to A in a unique way we will demonstrate in the following section, that Equation (22) gives a good enough estimate of A.

Comparison with Simulations by Limbach and Holm

Limbach and Holm[22] performed molecular dynamics simulations with polyelectrolytes at variable solvent quality, strength of electrostatic interaction and charge fraction of the polyelectrolytes. Variation of these properties changes the extent of counterion condensation and thus, the effective charge density of the fully dissociated polyelectrolyte chains. The effective charge of the chains, in turn, determines the size and shape adopted by the chains. Although simulations were performed at a poor solvent quality for the chain backbone, necklace like shapes for the shrinking chains could only be observed in a narrow regime of charge fraction and of strength of electrostatic interaction.

Scattering curves of those intermediates are suitable for a comparison with our model. Their validity lies in the exact knowledge of the corresponding number of pearls, mean pearl size and distance between two neighboring pearls, which are summarized in Table 1. Two intermediate structures were selected: One with an ensemble average number of 1.87 pearls per polyelectrolyte chain and one with an ensemble average number of 2.9 pearls per polyelectrolyte chain. Scattering curves of both selected intermediates were represented in Figure 25 of Ref. 22. The selected curves were compared with scattering curves from our pearl necklace model. Translation of the simulations to an actual length scale in our model was performed by setting $b = 2.591$ Å$^{-1} = \sigma$ with σ the unit length used in Ref. 22.

© 2004 WILEY-VCH Verlag GmbH & KGaA, Weinheim

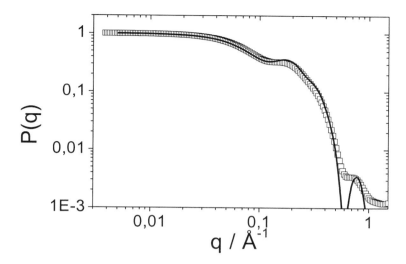

Figure 6. Form factor of an intermediate structure with an ensemble average number of 1.87 pearls per polyelectrolyte chain, generated by molecular dynamics simulations[22] (□) in comparison with the form factor of a dumbbell (—) calculated by Equation (12) with A=40 Å and R=7.5 Å . Parameters are summarized in Table 1.

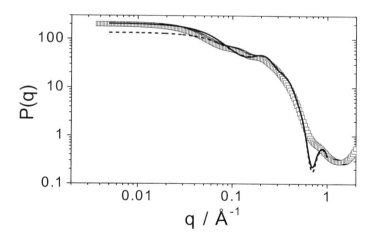

Figure 7. Scattering curve of an intermediate structure with an ensemble average number of 2.9 pearls per polyelectrolyte chain, generated by molecular dynamics simulations[22] (□) in comparison with model scattering curves calculated by Equation (12): Trimbell (—) with A=35 Å   and R=6.5 Å; (---) dumbbell with A=35 Å   and R=6.5 Å. All curves are unnormalized and approach M_w in the limit of q=0. Parameters are summarized in Table 1.

© 2004 WILEY-VCH Verlag GmbH & KGaA, Weinheim

With both curves taken from Ref. 22, a value of A and R could be estimated by means of Equation (22) and Equation (23) and compared to the real values known from the simulated samples.[22] As is demonstrated in Table1, agreement is satisfactory. By inserting these estimates for A and R into Equation (12), model curves were calculated for a dumbbell and a trimpbell (necklace with three pearls).

Complete description of the scattering of the simulated dumbbell could be achieved. Results are shown in Figure 6. In the case of the trimpbell, the model curve is adapted to the simulated curve in a regime of $q > 0.12$ Å^{-1}. This was achieved by using the second inflection point to estimate A according to Equation (22). Satisfactory agreement is observed only for $q > 0.12$ Å^{-1} corresponding to the regime, which is dominated by the element of a dumbbell. Agreement is getting poorer for lower q-values. This can clearly be attributed to the stretched shape of the simulated chain, which causes interference effects between two pearls separated by two strings. Such interferences do not occur in our model, which is based on a freely jointed chain. Details can be taken from Figure 7. The very failure is also responsible for the larger radius of gyration R_g observed for the simulated trimpbell (Table 1).

Table 1: Comparison of pearl necklace parameters from molecular dynamics simulations[22] with the corresponding parameters from the theoretical curves for the present pearl necklace model shown in Figure 6 and 7. The parameters corresponding to the model curves are estimates based on an application of Equation (22) and Equation (23) to the scattering curves from molecular dynamics simulation.

Simulation by Limbach and Holm[22]				Pearl necklace model			
n_p	A Å	$R^\S$ Å	R_g Å	n_p	A Å	R Å	R_g Å
1.87	38.9	8.7	19.1	2	40	7.5	20.4
2.90	36.0	7.1	29.7	3	35	6.5	23.0

§Estimated from the average radius of gyration of a pearl devided by 0.78

However, proper estimates of R and A could be extracted even in the case of the trimpbell. We simply took advantage of the fact that a q-regime exists, which is dominated by the element of a dumbbell, irrespective of the size of the pearl necklace. This is illustrated in Figure 7 where unnormalized scattering curves were used. The theoretical curves of a trimpbell and a dumbbell, which have the same pearl size R and pearl distance A are compared. Clearly, the simulated trimpbell and the trimpbell and dumbbell based on the freely jointed chain show the same features if $q > 0.12$ Å^{-1} , i. e. a shoulder/minimum at $q_{min} = 0.15$ Å^{-1} and a shoulder/minimum at $q_s = 0.7$ Å^{-1}.

© 2004 WILEY-VCH Verlag GmbH & KGaA, Weinheim

Conclusions

Particle scattering curves could be derived for pearl necklace chains based on a freely jointed chain. The chains consist of two components, homogeneous spheres and infinitely thin rods. The pearls are connected by the rod like segments which have a constant length. Both components contribute to the scattering of the pearl necklace.

The scattering curves have a characteristic shoulder or minimum. Its location correlates with the distance between neighboring pearls. For scattering vectors larger than this location, the scattering behavior is dominated by the individual pearls.

Model scattering curves were compared with molecular dynamics simulations performed by Limbach and Holm.[22] A successful description of simulated data with model curves was only successful in the case of a dumbbell, which was the most simple case. In the case of a trimpbell, the model could not reproduce correlations which stem from pearls separated by two rods. As already pointed out by Limbach and Holm,[22] simulated pearl necklace intermediates generated from fully dissociated polyelectrolyte chains by simple counter ion condensation are highly stretched entities. For this type of shrinking process, the model of Rubinstein et al.[17] is superior to our freely jointed pearl necklace.

Irrespective of the degree of flexibility of a pearl necklace, the structural element of a dumbbell plays a central role in the proof of a pearl necklace shape. A correlation between neighboring pearls and intra-spherical interferences are the most frequent features in the scattering curve. Once these features become discernible in a scattering experiment, they give access to A and R by simply mapping the appropriate section of the experimental curve to the dumbbell scattering curve.

Acknowledgement. Financial support of the Deutsche Forschungsgemeinschaft, Schwerpunktprogramm "Polyelektrolyte mit definierter Molekülarchitektur" SPP 1009, is gratefully acknowledged. The authors are indebted to H. J. Limbach, Ch. Holm and K. Kremer for making available the data of two scattering curves including helpful comments.

© 2004 WILEY-VCH Verlag GmbH & KGaA, Weinheim

42

[1] F. T. Wall, J. W. Drenan, *J. Polym. Sci.* **1951**, *7*, 83.
[2] I. Michaeli, *J. Polym. Sci.* **1960**, *48*, 291.
[3] A. Ikegami, N. Imai, *J. Polym. Sci.* **1962**, *56*, 133.
[4] H. Eisenberg, G. R. Mohan, *J. Phys. Chem.* **1959**, *63*, 671.
[5] H. Eisenberg, E. F. Casassa, *J. Polym. Sci.* **1960**, *47*, 29.
[6] K.Huber, *J. Phys. Chem.* **1993**, *97*, 9825.
[7] R. Schweins, K. Huber, *Eur. Phys. J. E* **2001**, *5*, 117.
[8] M. Meewes, J. Ricka, M. de Silva, R. Nyffenegger, Th. Binkert, *Macromolecules* **1991**, *24*, 5811 and references therein.
[9] C. Wu, S. Zhou, *Macromolecules* **1995**, *28*, 5388.
[10] C. Wu, S. Zhou, *Macromolecules* **1995**, *28*, 8381.
[11] X. Wang, X. Qiu, C. Wu, *Macromolecules* **1998**, *31*, 2972.
[12] B. Chu, Q. Ying, A. Y. Grosberg, *Macromolecules* **1995**, *28*, 180.
[13] B. Chu, Q. Ying, *Macromolecules* **1996**, *29*, 1824.
[14] A. R. Khokhlov, *J. Phys. A: Math. Gen.* **1980**, *13*, 979.
[15] Y. Kantor, M. Kardar, *Europhys. Lett.* **1994**, *27*, 643.
[16] Y. Kantor, M.Kardar, *Phys. Rev. E* **1995**, *51*, 1299.
[17] A. V. Dobrynin, M. Rubinstein, S. P. Obukhov, *Macromolecules* **1996**, *29*, 2974.
[18] F. J. Solis, M. Olvera de la Cruz, *Macromolecules* **1998**, *31*, 5502.
[19] U. Micka, Ch. Holm, K. Kremer, *Langmuir* **1999**, *15*, 4033.
[20] P. Chodanowski, S. Stoll, *J. Chem. Phys.* **1999**, *111*, 6069.
[21] A. V. Lyulin, B. Dünweg, O. V. Borisov, A. A. Darinskii, *Macromolecules* **1999**, *32*, 3264.
[22] H. J. Limbach, Ch. Holm, *J. Phys. Chem. B* **2003**, accepted.
[23] V. O. Aseyev, S. I. Klenin, H. Tenhu, I. Grillo, E. Geissler, *Macromolecules* **2001**, *34*, 3706.
[24] M.-J. Li, M. M. Green, H. Morawetz, *Macromolecules* **2002**, *35*, 4216.
[25] A. Kiriy, G. Gorodyska, S. Minko, W. Jaeger, P. Štepánek, M. Stamm, *J. Am. Chem. Soc.* **2002**, *124*, 13454.
[26] R. Schweins, P. Lindner, K. Huber, *Macromolecules* **2003**, submitted.
[27] R. Schweins, G. Goerigk, M. Ballauff, K. Huber, in preparation.
[28] C. Heitz, M. Rawiso, J. Francois, *Polymer*, **1999**, *40*, 1637.
[29] Yu. A. Kuznetzov, E. G. Timoshenko, K. A. Dawson, *J. Chem. Phys.* **1995**, *103*, 4807 and *J. Chem. Phys.* **1996**, *104*, 3338.
[30] P. Debye, *Ann. Phys. Leipzig* **1915**, *46*, 809.
[31] W. Burchard, K. Kajiwara, *Proc. Roy. Soc. Lond.* **1970**, *A 316*, 185.
[32] J. Hermans, J. J. Hermans, *J. Phys. Chem.* **1958**, *62*, 1543.
[33] S. Chandrasekhar, *Rev. Mod. Phys.* **1943**, *15*, 1.
[34] Rayleigh, Lord *Proc. Roy. Soc.* **1914**, *A 90*, 219.
[35] T. Neugebauer, *Annalen der Physik* **1943**, *42*, 509.
[36] R. Schweins, K. Huber, *Polymer* **2003**, submitted.

© 2004 WILEY-VCH Verlag GmbH & KGaA, Weinheim

Macromol. Symp. **2004**, *211*, 43-53

Conformations and Solution Structure of Polyelectrolytes in Poor Solvent

Hans Jörg Limbach, Christian Holm,* *Kurt Kremer*

Max-Planck-Institut für Polymerforschung, Ackermannweg 10, 55128 Mainz, Germany; Fax: (+49) 6131 379100; Email: {limbach, holm, kremer}@mpip-mainz.mpg.de

Summary: Using extensive Molecular Dynamics (MD) simulations we study the behavior of polyelectrolytes in poor solvents, where we take explicitly care of the counterions. The resulting pearl-necklace structures are subject to strong conformational fluctuations, only leading to small signatures in the form factor, which is a severe obstacle for experimental observations. In addition we study how the necklace collapses as a function of Bjerrum length. At last we demonstrate that the position of the first peak in the inter-chain structure factor varies with the monomer density close to $\rho_m^{1/3}$ for all densities. This is in strong contrast to polyelectrolyte solutions in good solvent.

Keywords: polyelectrolytes, hydrogels, computer modeling, molecular dynamics, solution properties

Introduction

Polyelectrolytes (PEs) are polymers which have the ability to dissociate charges in polar solvents resulting in charged polymer chains (macroion) and mobile counterions. They represent a broad and interesting class of soft matter [1, 2] that enjoy an increasing attention in the scientific community. In technical applications PEs are used as viscosity modifiers, as precipitating agents, and as superabsorbers. A thorough un-

© 2004 WILEY-VCH Verlag GmbH & KGaA, Weinheim

DOI: 10.1002/masy.200450703

derstanding of charged soft matter has become of great interest also in biochemistry and molecular biology. This is due to the fact that virtually all proteins, as well as other biopolymers such as DNA, actin, or microtubules are PEs.

Many PEs have a hydrocarbon based backbone for which water is a very poor solvent. Therefore, in aqueous solution, there is a competition between solvent quality, Coulombic interaction, and entropic degrees of freedom. The conformation of individual chains can under certain conditions assume pearl-necklace like structures[3, 4, 5]. In earlier simulations[6, 7] we found that the polymer density can be used as a very simple parameter to separate different conformational regimes. Here we analyze in more detail the single chain behavior and the scaling of the peak in the inter-chain structure function.

Simulation Model

Our PE model and molecular dynamics (MD) approach has been described in Refs. [6, 7, 8, 9, 10] and consists of a bead spring chain of Lennard-Jones (LJ) particles. Chain monomers interact via a LJ interaction up to a distance $R_c = 2.5\sigma$ and experience an attraction with $\epsilon = 1.75\,k_BT$. The Θ-point for this model is at $\epsilon = 0.34 k_B T$ [6]. The counterions interact via a purely repulsive LJ interaction. For bonded monomers we add a FENE bond potential. Charged particles at separation r interact via the Coulomb energy $k_BT\ell_b q_i q_j/r$, with $q_i = 1, (-1)$ for the charged chain monomers (counterions) and the Bjerrum length $\ell_b = e^2/(4\pi\epsilon_S\epsilon_0 k_BT)$ (e: unit charge, ϵ_0 and ϵ_S: permittivity of the vacuum and of the solvent). We simulated various systems with several chains in the central simulation box at various monomer densities ρ_m and different values of ℓ_b. Each chain consists of $N_m = 48\ldots478$ monomers, with a

© 2004 WILEY-VCH Verlag GmbH & KGaA, Weinheim

charge fraction $f = 1 \ldots 1/3$. The pressure p was found to be always positive, with the pV diagram being convex at all densities, thus our simulations are stable, reach true thermal equilibrium, and reside in a one phase region.

Single Chain Properties

With the help of a specially developed cluster algorithm[8, 10] that automatically recognizes the number of pearls in a conformation, we have analyzed all equilibrium conformations in our systems. We found large coexistence regimes between structures consisting of conformations with different pearl numbers. Even a single chain shows over the course of time many transitions between different pearl structures, hence the different pearl states are not frozen or metastable. Also the position and size of the pearls and strings is constantly changing[8, 9], compare Fig.1. We furthermore found, that the lower the pearl number, the stronger the counterions are attracted to the pearls. This is easy to rationalize, since smaller number of pearls mean larger pearls due to mass conservation, and thus to a higher local charge density. The integrated ion distribution versus the chain distance displays an inflection point, which is a signal of counterion condensation[11]. In contrast to analytical theories[12], the pearl structures are stable, even though there are counterions localized near and/or inside the pearls.

When one starts in a necklace conformation and increases ℓ_b the counterions will get attracted more and more towards the chain. Scaling theories have predicted that with the onset of condensation the necklace state should collapse in a first order transition into the globular state[13, 5, 12, 14]. However, the 'onset' of condensation is not a sharp border, rather like within Poisson-Boltzmann theory one finds a smooth

© 2004 WILEY-VCH Verlag GmbH & KGaA, Weinheim

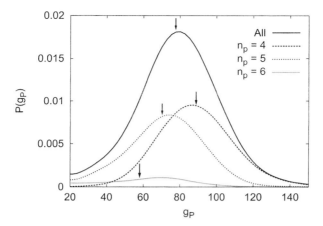

Figure 1: Probability distributions P for the pearl size g_P for a system with chain length $N_m = 382$, $\ell_b = 1.5\sigma$, and $f = 1/3$. Shown is the distribution for all chains as well as the distributions for the different structure types. The arrows mark the mean value of the corresponding probability distribution.

distribution of counterions which gets weighted closer to the macroion as the coupling is increased[11]. Accordingly, in the simulation we do not observe a collapse transition. The picture is qualitatively the same as in the good solvent case[15]. At $\ell_b = 0$ (no electrostatic interaction) the chain is in a collapsed conformation. By increasing ℓ_b the chain first extends up to a maximum, and then slowly shrinks back to a collapsed state. The non-monotonic behavior of the extension is qualitative the same as in the good solvent case[15] however the decrease is faster and more pronounced here[6, 16]. There is also a subtle dependence on f. The scaling variable which determines R_E of the necklace is $f^2\ell_b$ at fixed N_m and ϵ[5]. In Fig.2 we show snapshots of chains with

© 2004 WILEY-VCH Verlag GmbH & KGaA, Weinheim

the same value of $f^2\ell_b$, but different f. The chain extension is drastically different, and depends on the local interactions mediated by the counterions. This effect is obviously not captured by the scaling Ansatz! Note also, that the conformations on the way to the collapsed globule are very much reminiscent of a cylindrical shape[13], since the strings become very short, and the pearls coalesce slowly on a sausage like string until they reach the globular state. The collapsed state is reached roughly at the same value of ℓ_b, which is reminiscent of the critical point of a Coulomb fluid.

Next we computed the spherically averaged form factor $S_1(q)$ of a single chain, shown in Fig.3, since this is an observable that is accessible in experiments, and for which also theoretical expressions have been developed[5, 17, 18].

In the range $1 < q\sigma < 2$ we denote a sharp decrease in S_1, which reflects the intra pearl scattering, because it shows the typical Porod scattering of $S_1(q) \simeq q^{-4}$. The kink at $q\sigma \approx 1.66$ appears at the position expected from the pearl size, but is broadly smeared out due to large size fluctuations. The shoulder which can be seen at $q\sigma \approx 0.5$ does not come from the intra-pearl scattering but is due to the scattering of neighboring pearls along the chain (inter-pearl contribution), which have a mean distance of $\langle r_{PP} \rangle = 13.3.\sigma$. It is also smeared out due to the large distribution of inter-pearl distances. We conclude that the signatures of the pearl-necklaces are weak already for monodisperse samples. A possible improvement could be achieved for chains of very large molecular weights and low pearl numbers, which could lead to stable and large signatures.

© 2004 WILEY-VCH Verlag GmbH & KGaA, Weinheim

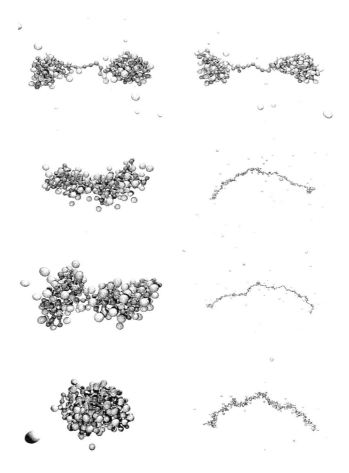

Figure 2: Snapshots for different values of the scaling variable $f^2\ell_b$. Left row with $f = 1/3$, right row with $f = 1/2$. From top to bottom $f^2\ell_b$ has the values: 0.08σ, 0.25σ, 0.5σ, 1.0σ. System: 8 chains with $N_m = 199$ monomers at $\rho_m = 5.0 \times 10^{-5}\sigma^{-3}$.

© 2004 WILEY-VCH Verlag GmbH & KGaA, Weinheim

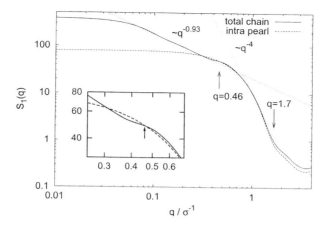

Figure 3: Spherically averaged form-factor $S_1(q)$. Shown are the single chain form-factor (solid line), together with the part of the form factor coming from the intra-pearl scattering (dashed line). The dotted and short dashed fits show the elongated chain part, and the Porod scattering part (globular conformation). System: 7 chains with 382 monomers, $f = 1/3$, $\ell_b = 1.5\sigma$, $\rho_m = 1.0 \times 10^{-5}\sigma^{-3}$.

Scaling of the Correlation Length ξ in Solution

The overall scattering function $S(q)$ of the solution contains additional experimental information. For good solvent PEs, experiments [19], theory [20], and simulations [15] find a pronounced first peak of $S_{IC} = S/S_1$ at $q^* = (2\pi)/\xi$, where ξ is the correlation length. The position varies as $q^* \propto \rho_m^{1/3}$ in the very dilute regime and crosses over to a $\rho_m^{1/2}$ regime at higher concentrations. In Fig. 4 we have plotted the density dependence of q^* in poor solvent for different chain lengths. Within the error bars we find that for poor solvent chains q^* scales proportional to $\rho_m^{0.35\pm0.04}$ for *all* concentrations and

© 2004 WILEY-VCH Verlag GmbH & KGaA, Weinheim

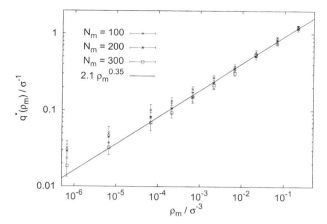

Figure 4: Density dependence of the peak q^* in the structure factor for three different chain length $N_m = 100, 200, 300$ with $f = 0.5$ and $\ell_b = 1.5\sigma$. The black line is a fit to the data with $N_m = 200$.

chain lengths. This is in accord with very recent experiments[21], but theoretically not well understood. The response of the PE conformation to density changes is much larger in the poor solvent case [6, 8] than in the good solvent case [15], and the chain extension behaves non-monotonic as a function of density [6, 8]. Furthermore, in the density regime between $\rho_m\sigma^3 = 10^{-2} \ldots 10^{-4}$ the chain extension and the pearl number varies most strongly, and almost all monomers are located within the pearls. Upon approaching the dense regime, the string length tends to zero and we find a chain of touching pearls, indicating that the conventional necklace picture breaks down. Our result is compatible to scaling exponents found in scattering experiments [17, 22, 23].

Scaling theories[24, 25] have predicted a $\rho_m^{1/2}$ regime to start at ρ_o^*, which is defined

© 2004 WILEY-VCH Verlag GmbH & KGaA, Weinheim

at the density where $R_E \approx \xi$, and to extend until $\xi \approx r_{pp}$ where a bead-controlled $\rho_m^{1/3}$ regime starts. We find $\rho_o^* \sigma^3 \simeq 5 \times 10^{-2}, 10^{-3}, 10^{-4}$ for $N_m = 100, 200, 300$. A pearl-pearl separation of the order of the correlation length length, $r_{pp} \approx \xi$, is reached between $\rho_m \sigma^3 = 10^{-2}$ and 10^{-1}, which is roughly independent of N. Especially for the longer chains ($N_m = 200, 300$) a clear signature of a different power law, i.e. $\rho_m^{1/2}$, should be visible. One possible reason for our different findings is that the strong inter-chain coupling and the influence of the counterions on the conformations are not sufficiently taken into account in the (mean field) scaling approach. It is not clear at this stage if the $\rho_m^{1/2}$ regime can be recovered for much longer chain length. In addition we observe that the chains form a transient physical network at $\rho_m \sigma^3 = 0.2$ for $N_m \geq 200$ which has neither been seen in previous simulations nor predicted by theoretical approaches but is in accord with experimental studies [17, 22, 23]. During the simulation time these networks reconstruct several times, e.g. chains are not trapped!

A more detailed account of the presented material will be published in forthcoming publications[10, 26].

Acknowledgments

We gratefully acknowledge partial funding through the "Zentrum für Multifunktionelle Werkstoffe und Miniaturisierte Funktionseinheiten", grant BMBF 03N 6500, and the DFG through the SFB 625 and the TR 6.

© 2004 WILEY-VCH Verlag GmbH & KGaA, Weinheim

[1] *Polyelectrolytes: Science and Technology*, M. Hara Ed., Marcel Dekker, New York, 1993.

[2] *Electrostatic Effects in Soft Matter and Biophysics*, Vol. 46 of *NATO Science Series II - Mathematics, Physics and Chemistry*, C. Holm, P. Kékicheff, and R. Podgornik, Eds. Kluwer Academic Publishers, Dordrecht, NL, 2001.

[3] Y. Kantor, M. Kardar, Europhys. Lett. **1994**, *27*, 643 .

[4] Y. Kantor, M. Kardar, Phys. Rev. E **1995**, *51*, 1299.

[5] A. V. Dobrynin, M. Rubinstein, S. P. Obukhov, Macromolecules **1996**, *29*, 2974.

[6] U. Micka, C. Holm, K. Kremer, Langmuir **1999**, *15*, 4033.

[7] H. J. Limbach, C. Holm, J. Chem. Phys. **2001**, *114*, 9674.

[8] H. J. Limbach, Ph.D. thesis, Johannes Gutenberg Universität, Mainz, Germany, 2001, available at http://archimed.uni-mainz.de/pub/2002/0121/

[9] H. J. Limbach C. Holm, Comp. Phys. Comp. **2002** *147*, 321; H. J. Limbach, C. Holm, K. Kremer, Europhys. Lett. **2002** *60*, 566.

[10] H. J. Limbach C. Holm, J. Phys. Chem B **2003**, *107*, 8041.

[11] M. Deserno, C. Holm, S. May, Macromolecules **2000**, *33*, 199.

[12] H. Schiessel P. Pincus, Macromolecules **1998**, *31*, 7953.

[13] A. Khokhlov, J. Phys. A **1980**, *13*, 979.

[14] T. A. Vilgis, A. Johner, J.-F. Joanny, Eur. Phys. J. E **2000**, *2*, 289.

[15] M. J. Stevens, K. Kremer, J. Chem. Phys. **1995**, *103*, 1669.

[16] M. Khan, S. Mel'nikov, B. Jönsson, Macromol. **1999**, *32*, 8836.

[17] C. Heitz, M. Rawiso, J. François, Polymer **1999**, *40*, 1637.

[18] R. Schweins, K. Huber, preprint; see also contribution in this volume.

[19] M. Nierlich *et al.*, J. Physique **1979**, *40*, 701.

[20] J. F. Joanny, in *Electrostatic Effects in Soft Matter and Biophysics*, Vol. 46 of *NATO Science Series II - Mathematics, Physics and Chemistry*, C. Holm, P. Kékicheff, and R. Podgornik. Eds., Kluwer Academic Publishers, Dordrecht, NL, 2001, pp. 149–170.

© 2004 WILEY-VCH Verlag GmbH & KGaA, Weinheim

[21] D. Baigl , R. Ober, D. Quo, A. Fery, C. E. Williams, Europhys. Lett. **2003**, *62*, 588.

[22] W. Essafi, F. Lafuma, C. E. Williams, J. Phys. II **1995**, *5*, 1269.

[23] T. A. Waigh, R. Ober, C. E. Williams, J.-C. Galin, Macromolecules **2001**, *34*, 1973.

[24] A. V. Dobrynin, M. Rubinstein, Macromolecules **1999**, *32*, 915.

[25] A. V. Dobrynin, M. Rubinstein, Macromolecules **2001**, *34*, 1964.

[26] H. J. Limbach, C. Holm, K. Kremer, in preparation.

© 2004 WILEY-VCH Verlag GmbH & KGaA, Weinheim

Structure of Polyelectrolyte Solutions: Influence of Salt and Chain Flexibility

Roland G. Winkler

Institut für Festkörperforschung, Forschungszentrum Jülich, 52425 Jülich,
Germany
Email: r.winkler@fz-juelich.de

Summary: The structural properties of polyelectrolyte solutions with and without added salt are investigated using an integral equation theory approach. For
rodlike systems, all components of the solution (polyions, counterions, anions, and
cations) are treat explicitly. Solutions of flexible polyelectrolytes are investigated
using the Debye-Hückel model. The Polymer-Reference-Interaction-Site model for
the multicomponent system together with the Reference-Laria-Wu-Chandler closure
is solved numerically. It is show that the Coulomb interaction leads to a long-range
liquid like structure. Addition of salt causes screening of the interaction and the
structure disappears at very high salt concentrations. Comparison of the correlation
functions of the multicomponent systems with those obtained by a one-component
Debye-Hückel model exhibits good agreement. Moreover, for an appropriate range
of parameters, such as density or Bjerrum length, a shell of equally charged saltions
exists in the vicinity of a polyion. The effective potential between two monomers
displays attraction among the equally charged polyions. The strength of the attraction
increases with increasing salt concentration.

Keywords: correlation function; effective potential; polyelectrolytes; PRISM; salt

Introduction

Polyelectrolyte solutions have attracted considerable attention over the last few decades but they
are still one of the least understood colloidal systems both from the theoretical as well as experimental point of view. Understanding of the basic properties of such solutions is desirable, since
polyelectrolytes play a fundamental role in many industrial applications and they are elementary
for biological systems.[1–6] The coupling of various length scales introduced by the long-range nature of the Coulomb interaction, counterion condensation, and screening effects render polyelectrolytes difficult to study. Computer simulations provide valuable insight into the structural prop-

erties of polyelectrolyte solutions but require special techniques to treat the long-range Coulomb interaction, e.g., Ewald summation, and hence are often limited to small systems with short chains and/or low densities.[7–9] Liquid state theory based upon the Polymer-Reference-Interaction-Site model (PRISM)[10] offers another theoretical approach to polyelectrolyte solutions. PRISM theory is an extension of the Ornstein-Zernike equation[11] of atomic systems to molecular systems by taking the connectivity of the sites of a molecule into account by the chain structure factor.

Polyelectrolytes cover the whole range of stiffnesses from flexible to rodlike polymers depending on the chain length or the solution properties like the salt concentration. As a consequence, the structure of a solution and the conformations of the polyelectrolyte chains are strongly coupled. To account for this coupling, the solutions of the PRISM equations and the corresponding conformations of the polyelectrolytes have to be determined simultaneously. This is achieved by a recently developed extension of the PRISM theory.[12–22] Here, the non-pairwise intermolecular many chain interactions of a particular chain are cast into an effective pairwise intramolecular solvation potential which is determined self-consistently. This so called medium induced potential can be expressed by correlation functions calculated with the PRISM theory.[23–26]

In this paper we discuss the structural properties of polyelectrolyte solutions with and without added salt by investigating the various pair correlation functions for different values of important parameters such as Bjerrum length, monomer density, and/or salt density. In particular we will demonstrate that added salt leads to a screening of the Coulomb potential which is well captured by the Debye-Hückel potential. Furthermore, it will be shown that for large Bjerrum lengths and/or high salt densities a shell of oppositely charged saltions is found immediately outside the condensed counterion shell as it was already predicted theoretically for spherical polyions.[27] Moreover, we will demonstrate that addition of salt can lead to a stronger effective attraction between the monomers of two polyions than in the case of salt-free solutions.[28] Flexible chains in solution exhibit significant conformational changes with increasing density.[29] This leads to shifts in the characteristic peaks of the monomer-monomer pair correlation functions, particular at higher packing fractions.

© 2004 WILEY-VCH Verlag GmbH & KGaA, Weinheim

Prism and Model

The PRISM theory is a liquid state theory for molecular systems and can be obtained by an extension of the well-know Ornstein-Zernike equation[11] taking the connectivity of the chain molecules into account explicitly. The Ornstein-Zernike equation as well as the PRISM theory connect the total correlation function $h(r)$ with the so called direct correlation function $c(r)$ and, in case of molecular systems, with the intramolecular distribution function $\omega(r)$. The total correlation function $h(r)$ is related to the pair correlation function $g(r) = 1 + h(r)$ and the static structure factor $S(k)$ which can be expressed via $S(k) = \omega(k) + \rho h(k)$, where $\omega(k)$ is the intramolecular structure factor. In the case of the four-component system there are 16 different total correlation functions which will be denoted by $h_{ij}(r)$ $(i, j \in \{m, c, -, +\})$, where the index m refers to monomers of different polyions, c to counterions, $-$ to the negatively charged saltions, and finally $+$ to the positively charged saltions. For symmetry reasons, the correlation functions h_{ij} and h_{ji} are equal for all combinations of i and j, hence the number of different correlation functions immediately reduces to ten. The different correlation functions are all coupled by the PRISM equations which can be written conveniently in Fourier space as

$$\mathbf{h}(k) = \boldsymbol{\omega}(k)\mathbf{c}(k)\boldsymbol{\omega}(k) + \boldsymbol{\omega}(k)\mathbf{c}(k)\boldsymbol{\rho}\mathbf{h}(k) , \tag{1}$$

where $\mathbf{h}$ and $\mathbf{c}$ are the matrices containing the different total and direct correlation functions. The matrices $\boldsymbol{\omega}$ and $\boldsymbol{\rho}$, respectively, are composed of the intramolecular structure factors $\omega_{ij} = \omega_i \delta_{ij}$ and the various particle densities $\rho_{ij} = \rho_i \delta_{ij}$. In this formulation all monomers of a chain are considered to be equivalent, i.e., chain end effects are neglected, and hence only one monomer correlation function for the whole polyion has to be considered. Otherwise, if all monomers of the polyion would be treated explicitly, we would obtain a vast number of different correlation functions leading to intractable numerical problems especially for longer chains. The correlation functions can only be calculated when additional relations are provided between the direct correlation functions, the total correlation functions, and the intermolecular potential. In contrast to the PRISM equation (1) the so called closure relations cannot be calculated exactly for all systems. Therefore, many different closures, depending on the approximations, have been proposed in the literature.[10, 24, 30–32] For molecular systems with hard core interaction, the Reference-Laria-Wu-Chandler closure

$$\omega_i(r) * c_{ij}(r) * \omega_j(r) \;\; = \;\; \omega_i(r) * (c_{o,ij}(r) - \beta v_{ij}(r)) * \omega_j(r)$$

© 2004 WILEY-VCH Verlag GmbH & KGaA, Weinheim

$$+h_{ij}(r) - h_{o,ij}(r) \quad - \quad \ln\left(\frac{g_{ij}(r)}{g_{o,ij}(r)}\right) , \quad r > \sigma_{ij} ,$$

$$g_{ij}(r) = 0 , \quad r < \sigma_{ij} , \quad i,j \in \{m,c,+,-\} \tag{2}$$

proved to be very successful for the one- and two-component system.[33] Here, the index 0 denotes reference functions obtained for a pure hard core system at the same density using the PY closure. The asterisks denote convolution integrals and $\sigma_{ij} = (\sigma_i + \sigma_j)/2$ is the average of the diameters σ_i of the different particle species in the solution. When the intramolecular structure factors ω_i are known the PRISM equations (1) together with the closure relations (2) can easily be solved numerically using a Picard iteration scheme.[11]

For rodlike chains, the polyions are modelled as a linear sequence of N touching hard spheres of diameter σ_m and charge $Z_m e$. Since we assume that the polyions remain in a rodlike conformation for all parameter variations, the intramolecular structure factor is known and given by

$$\omega_m(k) = 1 + \frac{2}{N}\sum_{j=1}^{N-1}(N-j)\frac{\sin(jk\sigma_m)}{jk\sigma_m} . \tag{3}$$

The counterions and saltions are also modelled as charged hard spheres with diameters σ_c, σ_-, and σ_+ and charges $Z_c e$, $Z_- e$, and $Z_+ e$, respectively. Hence, the intramolecular structure factors of the counterions and saltions are all equal and given by $\omega_c(k) = \omega_+(k) = \omega_-(k) = 1$. Charge neutrality of the system relates the charges and densities of the different particle species with each other, i.e., $Z_m\rho_m + Z_c\rho_c + Z_+\rho_+ + Z_-\rho_- = 0$. Treating the solvent as a dielectric continuum with a dielectric constant ϵ, the intermolecular potential is given by

$$\beta v_{ij}(r) = \beta v_{ij}^{\mathrm{HC}}(r) + Z_i Z_j \frac{l_B}{r} ; \quad i,j \in \{m,c,+,-\} , \tag{4}$$

where $v_{ij}^{\mathrm{HC}}(r)$ is the hard core potential and $l_B = \beta e^2/\epsilon$ is the Bjerrum length ($\beta = 1/k_B T$).

The results presented in the following sections have been obtained for systems with monovalent ions, i.e., $Z_m = Z_+ = 1$ and $Z_c = Z_- = -1$. Furthermore, we assume that all ions are of the same size, i.e., we set $\sigma = \sigma_m = \sigma_c = \sigma_- = \sigma_+$. As a consequence, the negatively charged saltions and counterions are indistinguishable and the same correlation functions are obtained for these two species. Therefore, it is possible to reduce the four-component system to a three-component system containing only polyions, positively charged saltions, and negatively charged ions (counterions and negatively charged saltions) without any loss of information about the system.

© 2004 WILEY-VCH Verlag GmbH & KGaA, Weinheim

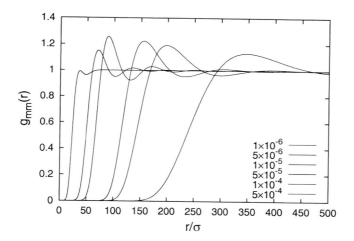

Figure 1. Monomer-monomer pair correlation functions $g_{mm}(r)$ for various monomer densities $\eta_m = \pi \rho_m \sigma^3/6$. The Bjerrum length is $l_B = 0.6\sigma$, the chain length is $N = 80$, and the salt densities are $\eta_+ = \eta_m$.

Correlation Functions of Rodlike Polyelectrolytes

Polyelectrolyte solutions exhibit a liquid-like order in dilute solution which diminishes at hight concentrations (cf. Figure 1). At infinite dilution $g_{mm}(r)$ has a value close to zero at small separations and monotonically increases to its asymptotic value of one at large separations. At low concentrations of polymer and salt, a peak appears on the length scale which is determined by the polyelectrolyte density. As the concentration is increased further, the liquid-like order first becomes more pronounced and then disappears at sufficiently high concentrations. This behavior has been obtained before by PRISM[33] and has been confirmed by computer simulations.[34] At fixed polymer concentration the structure disappears with increasing salt concentration as shown in Figure 2. However, significant changes are observed only, when the salt concentration exceeds the monomer concentration. For salt concentrations $\rho_+ > \rho_m$, further addition of salt leads to a strong screening of the Coulomb interaction, where the screening length decreases with increasing salt concentration, and the peak in the correlation function shifts to smaller length scales. Thus, the structure of the solution is no longer solely dependent on the polyelectrolyte concentration for salt concentrations exceeding such values.

© 2004 WILEY-VCH Verlag GmbH & KGaA, Weinheim

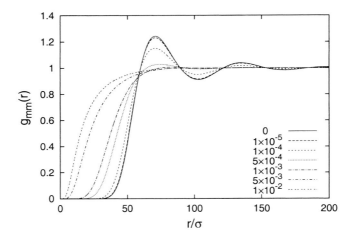

Figure 2. Monomer-monomer pair correlation functions $g_{mm}(r)$ for various salt concentrations η_+. The Bjerrum length is $l_B = 0.6\sigma$, the chain length is $N = 80$, and the monomer density is $\eta_m = 10^{-4}$.

The distribution of the counterions with respect to a polyion is described by the monomer-counterion correlation function $g_{mc}(r)$. As has been demonstrated for salt free solutions[17, 28] the counterions accumulate in the vicinity of a polyion at sufficiently high Bjerrum lengths, leading to a strongly peaked monomer-counterion correlation function at $r = \sigma$, whose height increases rapidly with the Bjerrum length. This also applies to the system with added salt, but there are pronounced quantitative differences. In the system with added salt the height of the peak of $g_{mc}(r)$ is much smaller than in the salt free system at the same Bjerrum length. A close examination shows that the peak height decreases when the salt concentration is increased. Since we interpret the large concentration of counterions close to a monomer as a manifestation of counterion condensation,[28] we find that the number of condensed counterions on the polyions is smaller than for systems without added salt. A similar effect, i.e., a decrease of the peak height in $g_{mc}(r)$, is achieved in salt free solutions by decreasing the Bjerrum length. Hence, addition of salt leads to a release (and/or exchange) of condensed counterions. Since counterions and negatively charged saltions are indistinguishable in our system, we have to take into account all negatively charged ions, and not only the counterions,

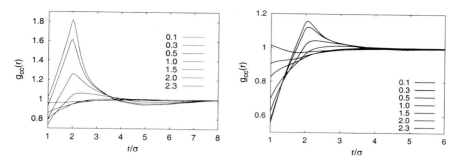

Figure 3. Counterion-counterion pair correlation functions for various Bjerrum lengths for a salt free system (left) and a system with added salt (right). The interaction strength decreases from top to bottom. The chain length is $N = 80$, the packing fraction of monomers $\eta_m = 10^{-2}$, and the packing fraction of saltions $\eta_+ = 10^{-2}$.

when we discuss counterion condensation. We then find that the number of condensed negatively charged ions in the system with added salt is actually higher than in a salt free solution. Thus, two effects contribute to screening in polyelectrolyte solutions: On the one hand screening is caused by changes of the salt density, which changes the Debye screening length. On the other hand, despite the decreasing correlation function with increasing salt concentration, the absolute amount of condensed ions increases with increasing salt concentration and the number of condensed charges in salt systems exceeds that of salt free systems. This has significant consequences on the monomer-monomer interaction.

A more detailed picture of counterion condensation is obtained from the counterion-counterion correlation function $g_{cc}(r)$. Counterion condensation causes the accumulation of counterions in the vicinity of the polyions. As a consequence, counterions closely approach each other despite the strong repulsive Coulomb interaction between them. Therefore, in salt free solution the counterion-counterion correlation function displays a peak at $r = 2\sigma$ for high Bjerrum lengths and the height of this peak increases rapidly with Bjerrum length[17,28] (cf. Figure 3). From Figure 3 we see that addition of salt leads to a decrease of the counterion-counterion correlation function. Hence, an increase of the salt concentration produces the same qualitative behavior as a decrease of the Bjerrum length in salt free solutions. With respect to condensed counterions, a decreasing correlation function corresponds to a decrease of the amount of condensed counterions. This is of course

© 2004 WILEY-VCH Verlag GmbH & KGaA, Weinheim

in perfect agreement with the decreasing peak height in g_{mc}, because less counterions near the polyion are equivalent with less counterions close to each other.

As already mentioned above, for the set of parameters used in this paper the counterions and negatively charged saltions are equivalent. Therefore all correlation functions regarding the negatively charged saltions are identical to those of the counterions, i.e., we have $g_{--} = g_{cc}$, $g_{m-} = g_{mc}$ and $g_{+-} = g_{c+}$. The remaining correlation function to be addressed is the correlation function between counterions and the positively charged saltions. Since the two species are oppositely charged, they closely approach each other for large Bjerrum lengths resulting in a peak of $g_{c+}(r)$ at $r = \sigma$ similar to the behavior found for $g_{mc}(r)$. Thus, not only counterions start to condense on the polyions for large Bjerrum lengths, but also positively charged saltions condense on the counterions. However, we find pronounced quantitative differences in the behavior of g_{mc} and g_{c+}. The peaks of the latter correlation function are much smaller for the same Bjerrum length. If we interpret this in terms of condensation, there are less positively charged saltions condensed on the counterions than counterions on the polyions for a given Bjerrum length. This behavior is in agreement with results found for counterion condensation in salt free solutions, where the number of condensed counterions per monomer decreases with decreasing chain length.[8, 28] When we consider a positively charged saltion as a polyion of chain length $N = 1$, it is perfectly clear why there are less counterions in the vicinity of saltions than in the vicinity of polyions.

Finally, we address the correlation function $g_{++}(r)$ between the positively charged saltions. As discussed above, for sufficiently large Bjerrum lengths the positively charged saltions condense on the counterions. Thus, there should be a shell of positively charged saltions in the vicinity of a polyion just outside the condensed counterions/saltions as already found theoretically for spherical polyions.[27] Figure 4 displays the correlation function $g_{++}(r)$ for the same set of parameters as in Figure 3. In contrast to Ref. [35], we indeed find a broad peak between $r = 3\sigma$ and $r = 4\sigma$, which increases with increasing Bjerrum length, indicating the saltion shell discussed above. In addition, we added the cation-cation correlation function for a monomeric system, i.e., a chain of length $N = 1$ at the same densities. It is obvious from Figure 4 that there is a pronounced chain length effect. With larger chain lengths more cations are condensed on the aggregate of polyion and counterions/anions than for shorter chains. (More details are presented in Ref. [36]).

© 2004 WILEY-VCH Verlag GmbH & KGaA, Weinheim

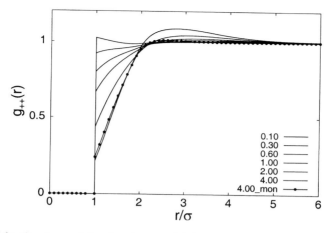

Figure 4. Salt-salt pair correlation functions $g_{++}(r)$ for various Bjerrum lengths l_B. l_B increases at $r = \sigma$ from top to bottom (parameters as in Figure 3). The line marked with dots corresponds to a solution of monomers (chain length $N = 1$) at the same densities.

Effective Potential for Rodlike Polyelectrolytes

Deeper insight into the consequences of counterion condensation is gained by an effective monomer-monomer and counterion-counterion potential, respectively.[28,37] The idea is to reduce the multicomponent system (macromolecules + counterions + salt) to effective one-component systems (macromolecules or counterions, respectively). We define the simplified model in such a way that the effective potential between the counterions or monomers, respectively, of the new system yields exactly the same correlation function (g_{cc}, g_{mm}) as found in the multicomponent case at the same density. Starting from the correlation function g_{cc} – respectively g_{mm} – of the multicomponent model we calculate an effective direct correlation function c_{eff} via the one-component Ornstein-Zernike equation. An effective potential is then obtained from the RLWC closures of the one- and multicomponent models.[28] For low and moderate densities the effective potential is well approximated by

$$\beta v_{ii,\text{eff}}(r) = \beta v_{ii}(r) + (c_{ii}(r) - c_{ii,\text{eff}}(r)) \ , \tag{5}$$

where $i \in \{m, c\}$. Hence, the effective potential is equal to the bare potential plus a modification given by the correlation functions of the multicomponent and one-component model.

© 2004 WILEY-VCH Verlag GmbH & KGaA, Weinheim

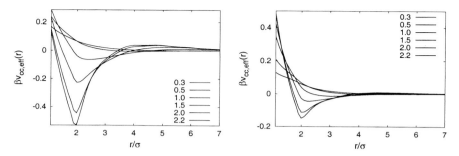

Figure 5. Effective potential between two counterions for various Bjerrum lengths for a salt free system (left) and a system with added salt (right). The interaction strength increases from top to bottom ($r \approx 2\sigma$). The chain length is $N = 80$, the packing fraction of monomers $\eta_m = 10^{-2}$, and the packing fraction of saltions $\eta_+ = 10^{-2}$.

Figure 5 (left) displays the effective counterion-counterion potential of a salt free system for various Bjerrum lengths. As is obvious from this figure, the effective potential is purely repulsive for low values of l_B and can very well be approximated by the bare Coulomb potential between the counterions. With increasing Bjerrum length the potential becomes negative for distances larger than a certain critical distance leading to an attractive force between two counterions. For even larger values of l_B, the effective potential exhibits a distinct minimum at a distance of about $r = 2\sigma$ in agreement with the position of the peak in g_{cc}. In systems with added salt, the qualitative structure of the effective counterion-counterion potential is similar to the salt free situation. However, the counterion attraction is weaker than in the salt free case. This can be considered a consequence of the screening of the Coulomb interaction among the counterions.

Similar to the effective counterion-counterion potential, we observe a minimum in the effective monomer-monomer potential at $r \approx 2\sigma$. Thus, also the oppositely charged polyions attract each other for sufficiently large interaction strengths. Comparing the potential for the salt free case (Figure 6 (left)) with the one of a system with added salt, we find a deeper minimum in the effective potential for the system with saltions, in contrast to the effective counterion potential. At a first glance, this behavior seems to contradict the results discussed in the previous section, which indicate a screening of the Coulomb interaction by addition of salt. Screening effects, however, take place on length scales larger than a few monomer diameters, whereas the attraction between

© 2004 WILEY-VCH Verlag GmbH & KGaA, Weinheim

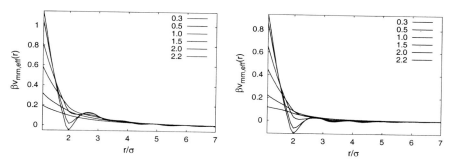

Figure 6. Effective potential between two monomers for various Bjerrum lengths for a salt free system (left) and a system with added salt (right). The interaction strength decreases from top to bottom ($r = \sigma$). The chain length is $N = 80$, the packing fraction of monomers $\eta_m = 10^{-2}$, and the packing fraction of saltions $\eta_+ = 10^{-2}$.

two polyions is only present on very short length scales.

The stronger attraction between the polyions in the system with added salt can be explained by two different effects. For the salt free case the attraction between two polyions is explained as a pure electrostatic effect, where the condensed counterions attract the oppositely charged monomers of another polyion by Coulomb interaction. As discussed in the previous section, for the system with added salt, there are in total more negatively charged ions adjacent to a polyion, leading to a stronger attractive interaction. On the other hand, the depletion interaction[38,39] has to be taken into account too. As is well known, depletion causes an attractive interaction between the large particles in a solution of small and large particles and the strength of the interaction increases with increasing density of the small particles. Since there are more negatively charged ions in the vicinity of a polyion in the system with added salt, the local density of ions in this region is higher than in the salt free case which may lead to a depletion interaction. The correlation functions for non-charged systems at the same densities display only a very weak depletion effect. Hence, the observed effect seems to be dominated by the Coulomb interaction.

It should be noted that for the parameters used in Figure 6 (right), the attraction between two polyions is so strong that it can even be seen in the monomer-monomer correlation function.[36] For large Bjerrum lengths a peak in $g_{mm}(r)$ appears at $r = 2\sigma$ in agreement with the position of the minimum in the effective potential. However, the peak cannot be observed in the monomer-

© 2004 WILEY-VCH Verlag GmbH & KGaA, Weinheim

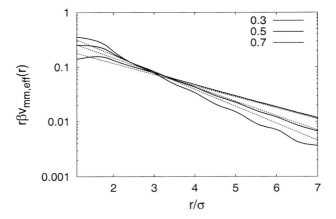

Figure 7. Scaled effective potential between two monomers for various Bjerrum lengths at the salt density $\eta_+ = 10^{-2}$. The interaction strength increases from top to bottom ($r = 7\sigma$). The chain length is $N = 80$ and the packing fraction of monomers $\eta_m = 10^{-2}$.

monomer structure factor $S_{mm}(k)$, where it should appear at $k\sigma \approx \pi$, hence the effect cannot be found in experiments measuring the structure factor.

The quantitative comparison of the obtained effective monomer-monomer potential with the Debye-Hückel potential exhibits excellent agreement for both, the dependence of the potential on the screening length as well as the dependence on the interaction strength. This applies for salt free systems[28] as well as for systems with added salt. Figure 7 shows the scaled effective potential for a system with added salt at various Bjerrum lengths. The dashed curves represent the Debye-Hückel potential $\beta v_{\mathrm{DH}} = l_B \exp\left(-\kappa r\right)/r$, where $\kappa = \sqrt{4\pi l_B [\rho_c Z_c^2 + 2\rho_+ Z_+^2]}$ is the inverse Debye screening length. Significant deviations form the Debye-Hückel potential are observed at monomer separations smaller than $r \approx 2\sigma$. This is not surprising, because at such small length scales the detailed shape of the interacting particles (monomers and counterions) matters. Naturally, further deviations are obtained at higher Bjerrum lengths due to counterion condensation.

© 2004 WILEY-VCH Verlag GmbH & KGaA, Weinheim

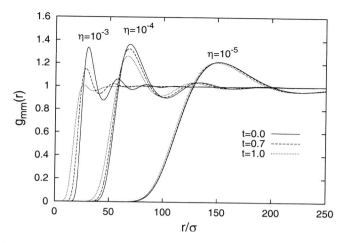

Figure 8. Monomer-monomer pair correlation functions of semiflexible chains of various stiffnesses t and densities η_m. The chain length is $N = 63$ and the Bjerrum length is $l_B = 0.5\sigma$.

Correlation Functions of Flexible Polyelectrolytes

To study the structural properties of solutions of flexible polyelectrolytes, we consider a one-component system of flexible chains interacting via the Debye-Hückel potential. As discussed in the last section, the Debye-Hückel potential provides an excellent description of multicomponent rodlike systems, thus, we expect it to capture the essential features of a system of flexible chains too. In studies of other systems the Debye-Hückel potential proved to be useful already.[33,40]

In order to adequately characterize the structural properties, possible changes in chain conformations have to be taken into account. As a consequence, the structure of a solution and the conformations of the polyelectrolyte chains are strongly coupled. To account for this coupling, the solutions of the PRISM equations and the corresponding conformations of the polyelectrolytes have to be determined simultaneously. This is achieved by a recently developed extension of the PRISM theory.[12–22] Here, the non-pairwise intermolecular many chain interactions of a particular chain are cast into an effective pairwise intramolecular solvation potential which is determined self-consistently. This so called medium induced potential (W) can be expressed by correlation functions calculated with the PRISM theory.[23–26] The total intramolecular potential V is then

© 2004 WILEY-VCH Verlag GmbH & KGaA, Weinheim

given by

$$V(\mathbf{r}) = \sum_{i=0}^{N} \sum_{j=i+1}^{N} \left[l_B \frac{e^{-\kappa|\mathbf{r}_i - \mathbf{r}_j|}}{|\mathbf{r}_i - \mathbf{r}_j|} + \beta W_{i,j}(|\mathbf{r}_i - \mathbf{r}_j|) \right] , \tag{6}$$

where $\beta W_{i,j}(k) = -\rho c(k)S(k)c(k)$ is the Fourier transformed medium induced potential. To analytically calculate the conformational properties of the semiflexible polyelectrolyte chain, we resort to the approximation scheme proposed by Edwards and Singh.[41] The details of the method are described in Ref. [29]. Here we will only present some of the results obtained by this approach.

Figure 8 depicts monomer pair correlation functions for various densities and chain stiffnesses ranging from flexible chains ($t = 0$) to rodlike chains ($t = 1$), respectively. The chain length is $N = 63$ and the Bjerrum length is $l_B = 0.5\sigma$. The semiflexible chains exhibit a similar liquid-like order as rodlike chains (cf. Figure 1). For the lowest density ($\eta = 10^{-5}$) the peak position and the peak height are almost independent from the chain stiffness despite the fact that the chain size is very sensitive to the stiffness in this density range as discussed in Ref. [29]. This is explained by the fact that the structure of the liquid in this density range is dominated by electrostatic interactions and not packing effects, which would depend on the chain size. With increasing density the peak sharpens and its height changes. Moreover, the peak height decreases with increasing stiffness. Therefore, an effect of the chain size should only be noticed for densities where packing effects have to be taken into account, i.e, for densities close and above to the overlap density. For flexible chains, the peak maximum is always located at larger distances than for stiffer chains. At the same density, the coils of the flexible chains exhibit a smaller overlap and hence a stronger Coulomb interaction. The comparison of the correlation functions for various Bjerrum lengths shows that the shift of the peaks and the increase of the peak height as a function of stiffness is less pronounced in the system with larger Bjerrum lengths, especially for low and moderate densities. This follows from the rodlike structure even of flexible chains at large Bjerrum lengths. Hence, the stiffness dependence of the correlation function disappears with increasing Bjerrum length. However, this applies only as long as the screening length, which increases with increasing density, is not too small. The chains behave like uncharged chains for small screening lengths and hence exhibit again a dependence on the chain stiffness.

© 2004 WILEY-VCH Verlag GmbH & KGaA, Weinheim

Conclusions

We have studied the structural properties of polyelectrolyte solutions composed of rodlike and flexible linear chain molecules, respectively. In particular, the influence of salt has been considered.

We find that addition of salt leads to a screening of the Coulomb interaction beyond a few monomer diameters. As a consequence, the characteristic peaks in the monomer-monomer correlation function decrease with increasing salt concentration and for very high salt concentrations g_{mm} is almost equivalent to the correlation function of an uncharged system. A detailed examination of the monomer-counterion and counterion-counterion correlation functions, respectively, reveals that there are less counterions condensed on a polyion, but the total number of adjacent negatively charged ions is actually larger than in the salt free case. We observe strong correlation effects among the charged particle next to a polyion. This is reflected by a shell of positively charged saltions formed around the condensed counterions at large Bjerrum lengths.

In addition, we have demonstrated that the various correlation functions can well be approximated by correlation functions of reduced systems containing a smaller number of components, e.g., only polyions, and using a Debye-Hückel potential with a salt density dependent screening length.

An extracted effective potential between polyions reflects the attractive interaction among the equally charged macroions for sufficiently large interaction strengths. Addition of salt leads to an enhancement of the attraction. For certain parameter combinations, the attraction is very strong and can even be detected in the monomer-monomer correlation function.

Chain flexibility influences the structural properties of a polyelectrolyte solution. Considering the pair correlation function, we find that semiflexible chains exhibit the same qualitative behavior as rodlike chains, but the characteristic peaks are shifted to larger distances and the height of the peaks increases with increasing density and decreasing chain stiffness due to conformational changes of the chains caused by the increasing density. This effect is more pronounced for lower densities and smaller Bjerrum lengths. There are various other aspects with respect of conformational changes of semiflexible chains, which have not been discussed in this article. Details of such systems can be found in Ref. [29].

Acknowledgement

The author gratefully acknowledges the financial support by the Deutsche Forschungsgemeinschaft within the Schwerpunktsprogramm 1009.

© 2004 WILEY-VCH Verlag GmbH & KGaA, Weinheim

[1] M. Mandel, *"Polyelectrolytes"*, D. Ridel, Dordrecht 1988.
[2] M. Hara, *"Polyelectrolytes: Science and Technology"*, Marcel Dekker, New York 1993.
[3] S. Förster, M. Schmidt, *Adv. Polym. Sci.* **1995**, *120*, 51.
[4] F. Buchholz, *Trends Polym. Sci.* **1982**, *99*, 277.
[5] A. Bhattacharya, *Prog. Polym. Sci.* **2000**, 25, 371.
[6] Y. Hayashi, M. Ullner, P. Linse, *J. Chem. Phys.* **2002**, 116, 6836.
[7] M. Stevens, K. Kremer, *J. Chem. Phys.* **1995**, 103, 1669.
[8] R. G. Winkler, M. Gold, P. Reineker, *Phys. Rev. Lett.* **1998**, 80, 3731.
[9] R. G. Winkler, M. O. Steinhauser, P. Reineker, *Phys. Rev. E* **2002**, 66, 021802.
[10] K. S. Schweizer, A. Yethiraj, *J. Chem. Phys.* **1993**, 98, 9053.
[11] J. P. Hansen, I. R. McDonald, *"Theory of Simple Liquids"*, Academic Press Limited, London 1986.
[12] J. Rudnick J. P. Donley, A. Liu, *Macromolecules* **1997**, 30, 1188.
[13] A. Yethiraj, K. S. Schweizer, *J. Chem. Phys.* **1992**, 97, 1455.
[14] K. S. Schweizer, K. G. Honnell, J. G. Curro, *J. Chem. Phys.* **1992**, 96, 3211.
[15] J. Melenkevitz, J. G. Curro, *J. Chem. Phys.* **1993**, 99, 5571.
[16] C. J. Grayce, A. Yethiraj, K. S. Schweizer, *J. Chem. Phys.* **1994**, 100, 6857.
[17] C. Y. Shew, A. Yethiraj, *J. Chem. Phys.* **1999**, 110, 5437.
[18] A. Yethiraj, *Phys. Rev. Lett.* **1997**, 78, 3789.
[19] A. Yethiraj, *J. Chem. Phys.* **1998**, 108, 1184.
[20] C. Y. Shew, A. Yethiraj, *J. Chem. Phys.* **2000**, 113, 8841.
[21] J. G. Curro, K. S. Schweizer, G. S. Grest, K. Kremer, *J. Chem. Phys.* **1989**, 91, 1357.
[22] K. S. Schweizer, J. G. Curro, *Adv. Chem. Phys.* **1997**, XCVIII, 1.
[23] D. Chandler, Y. Singh, D. M. Richardson, *J. Chem. Phys.* **1984**, 81, 1975.
[24] D. Laria, D. Wu, D. Chandler, *J. Chem. Phys.* **1991**, 95, 4444.
[25] J. Melenkevitz, K. S. Schweizer, J. G. Curro, *Macromolecules* **1993**, 26, 6190.
[26] Y. Singh, *J. Phys. A-Math Gen.* **1987**, 20, 3949.
[27] L. Belloni, *Chem. Phys.* **1985**, 99, 43.
[28] T. Hofmann, R. G. Winkler, P. Reineker, *J. Chem. Phys.* **2001**, 114, 10181.
[29] T. Hofmann, R. G. Winkler, P. Reineker, *J. Chem. Phys.* **2003**, 118, 6624.
[30] K. S. Schweizer, J. G. Curro, *Macromolecules* **1988**, 21, 3070.
[31] A. Yethiraj, C. K. Hall, *J. Chem. Phys.* **1990**, 93, 4453.
[32] A. Yethiraj, C. K. Hall, *J. Chem. Phys.* **1992**, 96, 7975.
[33] C. Y. Shew, A. Yethiraj, *J. Chem. Phys.* **1997**, 106, 5706.
[34] C. Y. Shew, A. Yethiraj, *J. Chem. Phys.* **1999**, 110, 11599.
[35] L. Harnau, P. Reineker, *J. Chem. Phys.* **2000**, 112, 437.
[36] T. Hofmann, R. G. Winkler, P. Reineker, *J. Chem. Phys.* **2003**, 119, 2406.
[37] M. Dymitrwska, Luc Belloni, *J. Chem. Phys.* **1999**, 111, 6633.
[38] A. Hanke, E. Eisenriegler, S. Dietrich, *Phys. Rev. E* **1999**, 59, 6853.
[39] A. A. Louis, E. Allahyarov, H. Löwen, R. Roth, *Phys. Rev. E* **2002**, *65*, 061407.
[40] M. Ullner, Woodward C. E, B. Jønsson, *J. Chem. Phys.* **1996**, 105, 2056.
[41] S. F. Edwards, P. Singh, *J. Chem. Soc., Faraday Trans. 2* **1979**, 75, 1001.

© 2004 WILEY-VCH Verlag GmbH & KGaA, Weinheim

Separation of Polyelectrolyte Chain Dynamics and Dynamics of Counterion Attachment by EPR Spectroscopy

Dariush Hinderberger, Hans Wolfgang Spiess, Gunnar Jeschke*

Max-Planck-Institut für Polymerforschung, Postfach 3148, 55021 Mainz, Germany
E-mail: jeschke@mpip-mainz.mpg.de

Summary: Rotational dynamics and local enrichment of counterions close to polyelectrolyte chains were studied by EPR spectroscopy in solvents of different viscosity. The results confirm previous findings (D. Hinderberger, G. Jeschke, and H. W. Spiess, Macromolecules 2002, 35, 9698) that electrostatic attachment of counterions to the chains is dynamic with lifetimes of contact ion pairs shorter than 1 ns. While in low-viscosity solvents linewidths for a dianionic nitroxide probe and their dependence on polyelectrolyte concentration are dominated by the gradient of local concentration in the vicinity of the chain, they are more strongly influenced by changes in rotational dynamics in a glycerol/water mixture. The slowdown of dynamics at higher viscosity strongly depends on polyelectrolyte concentration, suggesting that the lifetime of the attached state increases. The linewidths of trianionic triarylmethyl probes and of the center line of the nitroxide probes are dominated by local counterion enrichment both at low and high viscosity. Comparison of these linewidths and of the extent to which the lineshapes are non-Lorentzian indicates build-up of larger concentration gradients at higher viscosity.

Keywords: chain dynamics; counterion condensation; ESR/EPR; nanoheterogeneity; polyelectrolytes

Introduction

Polyelectrolytes – polymers with a large number of charged groups – play an important role in fields of scientific research as diverse as molecular biology and nanotechnology.[1,2] Furthermore, there is a wide variety of commercial applications already for polyelectrolyte materials in e.g. cosmetics, fuel cells, and food and oil industry.

Despite this widespread interest in and large amount of studies on polyelectrolytes, there are fundamental interactions whose influence on the polyelectrolyte structure has not yet been understood satisfactorily. Open questions include i) the role of conformation-dependent electrostatic interactions between charged repeat units of the polyelectrolyte on the one hand and between the polyelectrolyte and counterions on the other hand; ii) the role of especially *intra*molecular hydrophobic interactions, and iii) the role of specific solvation. It is commonly acknowledged that some of the very interesting properties of polyelectrolytes in

© 2004 WILEY-VCH Verlag GmbH & KGaA, Weinheim

DOI: 10.1002/masy.200450705

solution and at interfaces are a consequence of these interactions. Within the last decade theoretical investigations of the aforementioned interactions, driven by increasing computing power and sophisticated algorithms, have been carried out and led to predictions of polyelectrolyte conformations in solution that could not yet be verified experimentally.[3]

Many characterization methods can either only probe macroscopic properties (e.g. conductivity) or require some long-range order in the systems, such as the well-established scattering techniques (light-, x-ray-, and neutron-scattering). Magnetic resonance methods, such as electron paramagnetic resonance (EPR) spectroscopy,[4] being local, highly sensitive and highly selective, have potential to deliver valuable information on polyelectrolyte materials, which is otherwise inaccessible.

EPR spectroscopy in combination with spin labeling has been used for three decades to elucidate structure and dynamics of larger molecules, in particular of biomolecules (e.g. DNA, membrane proteins), but also in polymer science.[5,6] In spin labeling, the paramagnetic probe molecule is covalently attached to the molecule of interest, while in spin probing weak non-bonding interactions, such as hydrophobic interactions between polymers and spin-labeled surfactants are used to direct the label to the site of interest.[6] In polyelectrolyte-counterion systems spin probing can be based on the electrostatic interaction. In such experiments changes in the EPR spectra induced by changes in polyelectrolyte concentration are directly related to the interaction of the spin-carrying counterions with the polyelectrolyte chain.

Using divalent ionic spin probes together with continuous-wave (CW) EPR and Fourier-Transform (FT) EPR at room temperature we have recently shown that di- and trivalent probe ions are condensed to cationic polyelectrolyte chains via *dynamic electrostatic attachment*.[7] This term is used for counterion attachment in order to emphasize that a dynamic equilibrium between closely attached (site bound, according to Manning),[8] non-specifically attached (territorially bound) and detached (free) counterions is attained. Our previous work showed that exchange between site-bound and territorially bound spin-carrying counterions proceeds on a sub-nanosecond timescale.[7]

In our recent paper we moreover found that the effect of solvent in these systems is more complex than just that of a dielectric continuum between charges. Changing relative permittivity from $\varepsilon \approx 50$ (ethanol/water 1:1) to $\varepsilon \approx 140$ (N-methylpropionamide (NMPA)/water 2.3:1) did not result in a significant change of dynamic electrostatic attachment and local concentrations of probe ions, while the results for both these solvents differed strongly from the results for pure water. We attributed this to the interaction of the hydrophobic parts of the

© 2004 WILEY-VCH Verlag GmbH & KGaA, Weinheim

organic solvent molecules (in organic solvent/water mixtures) with the polyelectrolyte backbone, which leads to a screening of intra-polyelectrolyte hydrophobic interactions and thus prevents conformational changes that take place in pure water.

In this earlier study we did not discuss the effect of polyelectrolyte chain dynamics on the spectra. In our qualitative discussions we assumed that the polyelectrolyte molecules are fixed on the timescale of counterion dynamics. In this work we assess how the dynamics of counterion condensation is influenced by polyelectrolyte motion.

Our study is organized as follows. In the first part we describe new experimental results for a more viscous solvent mixture of water with glycerol and compare them with our earlier results.[7] In the second part, we extend our previous discussion of the effects taking particular care of the separation of spectral changes caused by a slowdown of rotational diffusion of the counterions due to temporary electrostatic attachment to the polyelectrolyte chain from those caused by more frequent counterion collisions due to increased concentration close to the polyelectrolyte chain. Finally, we consider how the two processes are influenced by polyelectrolyte chain dynamics.

Materials and Methods

As spin probes (Scheme 1) we used Fremy's salt dianion (Potassium nitrosodisulfonate), FS, with an unknown grade of purity (ICN Biomedicals) and triarylmethyl trianion, TAM (sodium salt, gift from Nycomed Innovations AB, Sweden). As cationic polyelectrolyte we chose poly(diallyldimethylammonium chloride), PDADMAC, with an M_W of 240 000, (Polysciences, Inc.). All chemicals were used as received. As solvent systems we chose deionized Milli-Q-water (permittivity: $\varepsilon_r = 80$ at 293 K), 66 wt.-% glycerol/34 wt.-% water (made from 87 wt.-% glycerol/water mixture, Fluka, density $\rho \approx 1.17$ g/ml, approximate $\varepsilon_r = 57$ at 293 K) and 70 vol.-% N-methylpropionamide (Aldrich Chem. Co.)/30 vol.-% water (NMPA/H_2O, $\rho = 0.95$ g/ml, approximate $\varepsilon_r = 140$ at 293 K). Small amounts of KOH were added to adjust the solution to pH > 8. Spin probe concentration was fixed at 0.5 mM (FS) or 0.4 mM (TAM), and polyelectrolyte concentration was varied from 4 mM (in monomeric units) to 140 mM. Data are presented as a function of the ratio R of spin probes to polymer repeat units, so that no assumption on the actual number of repeat units per chain, and thus on M_n, had to be made. EPR spectra at X-band (~9.7 GHz) were measured on a Bruker ELEXSYS 580 spectrometer using an AquaX inlet and a rectangular cavity (4103TM, Q-values typically ~3000). The temperature during all measurements was set to 293 K.

© 2004 WILEY-VCH Verlag GmbH & KGaA, Weinheim

Data analysis is based on three facts about the spectral changes observed on addition of polyelectrolyte that were established in our previous work.[7] First, there is line broadening for both the TAM and FS spin probes that affects all lines in the spectra to roughly the same extent and can be traced back to Heisenberg spin exchange broadening due to more frequent collisions of the counterion in regions close to the polyelectrolyte chain, where local concentration is enhanced. Second, this broadening is heterogeneous in the sense that a gradient of local concentration – from high spin probe concentration close the polyelectrolyte chains to approximately bulk concentration – leads to a distribution of relaxation times and thus to a non-Lorentzian lineshape. Third, for the FS spin probe additional broadening is observed that affects the high-field line to the largest and the center line to the smallest extent and can be traced back to a slowdown of rotational diffusion.

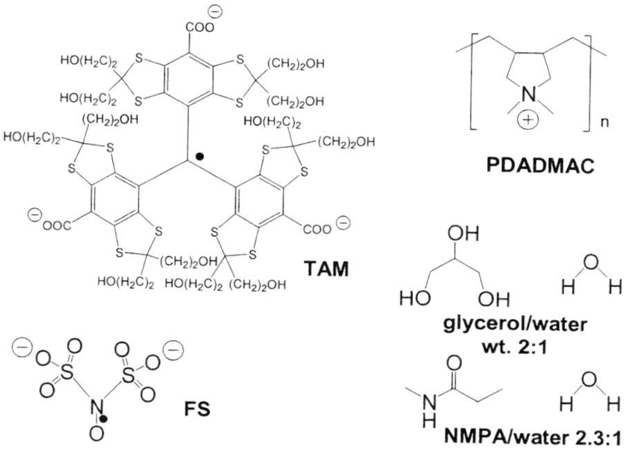

TAM = triarylmethyl trianion; FS = Fremy's salt dianion; PDADMAC = poly(diallyldimethylammonium chloride); NMPA = N-methylpropionamide.

Scheme 1. Molecular structures of charged spin probes, polyelectrolyte and solvents used in this study.

The EPR spectra of both spin probes throughout most of the polyelectrolyte concentration range can be fitted by a homewritten program which assumes that relaxation can be described by a stretched exponential decay.[9] The time-domain signal for a single line with resonance offset ω_i, characteristic transverse relaxation time T_2, and stretch factor x is then given by

$$V(t) = \sum_{k=1}^{n} \exp\left(-\frac{t}{T_{2,k}}\right)^{x} \exp(i\omega_k t), \qquad (1)$$

© 2004 WILEY-VCH Verlag GmbH & KGaA, Weinheim

where $n = 1$ for TAM and $n = 3$ for FS spin probes with $k = 1$ corresponding to the low-field line and $k = 3$ corresponding to the high-field line. The CW spectrum is obtained by Fourier-transformation of the sum signal $V(t)$, and subsequent pseudomodulation[10] with the same modulation amplitude as used in the experiments. This simulated CW EPR spectrum is fit to the experimental spectrum, which provides a quantification of the non-Lorentzian lines with a minimum number of fit parameters and without making any assumptions on the physical process leading to relaxation. It provides good fits of the data for all compositions of the system in low-viscosity solvents but is restricted to $R < 0.1$ in the glycerol/water mixture. Above this value we do not consider the simulated data reliable any more, as the deviation of the simulated spectra from the experimental data becomes significant.

All values reported for T_2 in this work are average values of the Williams-Watts distribution of T_2 [11]

$$\langle T_2 \rangle = \frac{1}{x} \cdot T_2 \cdot \Gamma\left(\frac{1}{x}\right), \tag{2}$$

where the brackets denote the average value and $\Gamma(1/x)$ is the gamma function.

A characteristic EPR linewidth $<\Delta B>$ of the non-Lorentzian lines can be computed by $\langle \Delta B_k \rangle = [0.412\ (G/MHz) * \langle T_{2,k} \rangle^{-1}]$.[12] This averaged linewidth is a quantitative measure of the electron spin-electron spin interaction: $<\Delta B> \propto \langle T_2 \rangle^{-1} \propto T_{2,intrinsic}^{-1} + T_{e-e}^{-1}$, which is in turn related to local concentration of the spin-carrying counterions. Here, $T_{2,intrinsic}^{-1}$ denotes the intrinsic relaxation rate at infinite dilution and T_{e-e}^{-1} denotes the concentration-dependent rate due to intermolecular electron spin interaction. The intrinsic rate differs for the three lines ($k = 1, 2, 3$) for nitroxides, and these differences are related to the rotational correlation time (see below). We assume broadening to be mainly due to Heisenberg spin exchange, but in solutions of high viscosity an additional effect due to residual magnetic dipolar interaction cannot be excluded.[13,14] As broadening due to both mechanisms is related to local concentrations our semi-quantitative conclusions do not depend on the detailed broadening mechanism.

For the FS spin probe, a characteristic value $\langle \tau_c \rangle$ for the rotational correlation time was computed from the so-called "B-parameter": $B = (<\Delta B(k=3)> - <\Delta B(k=1)>)/2$ as introduced by Goldman et al.[15, 16] Note that our $\langle \tau_c \rangle$ is not the strictly defined average of the distribution of rotational correlation times, which cannot be extracted by this simple data analysis procedure. This slightly modified data analysis compared to Ref. [7] does not lead to

© 2004 WILEY-VCH Verlag GmbH & KGaA, Weinheim

qualitatively different results but provides quantitative average values that are better characteristics of the distributions of T_2 and τ_c.

Results

Dynamic Electrostatic Attachment of Spin Probes

As reported earlier,[7] we have found that addition of oppositely charged polyelectrolyte to solutions of multivalent spin probes leads to a marked increase in the peak-to-peak linewidths and changes the lineshapes. Typical spectra for the probe FS in glycerol/water with and without added PDADMAC are displayed in Figure 1. Qualitatively, the lineshape changes on adding polyelectrolyte resemble those ones observed before in other solvent mixtures.[7]

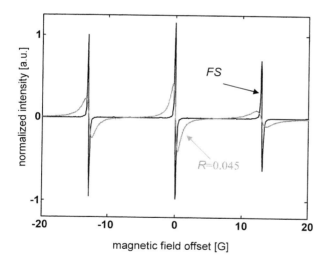

Figure 1. X-band (9.79 GHz) EPR spectra in 66 wt.-% glycerol/34 wt.-% water at two different spin probe/DADMAC ratios R. Black line: *no polyelectrolyte, only FS*; grey line: $R = 0.045$. Spectra were normalized to equal double integral.

However, the effects are much larger in glycerol/water than in low-viscosity solvents such as water and NMPA/water. By measurements of neutral and like charged nitroxide spin probes together with polyelectrolyte it has been verified that these changes are directly related to electrostatic interactions and are not pure viscosity effects.[7]

© 2004 WILEY-VCH Verlag GmbH & KGaA, Weinheim

Fremy's Salt and PDADMAC in Glycerol/Water and NMPA/Water

In Figure 2 parameters $\langle T_{2,k}\rangle$ and stretch factor x are shown for FS/PDADMAC in the low-viscosity solvent NMPA/water and the higher-viscosity solvent glycerol/water. Stretch factors x according to Equation (1) are a measure of the width of the distribution of T_2-relaxation times. With increasing R-value stretch factors deviate more and more strongly from their values in the absence of polyelectrolytes, thus the distribution of relaxation times becomes more heterogeneous with higher R. The trend for the stretch factors as a function of R is almost identical in both solvent systems, just the absolute values are shifted by approximately 0.05.

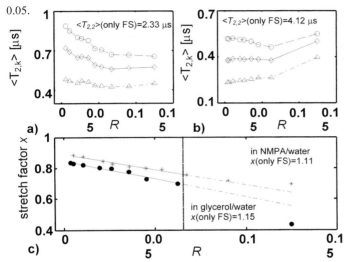

Figure 2. Relaxation time averages $\langle T_{2,k}\rangle$ and stretch factors x (see Equation (1) and (2)) for spin probe FS as function of FS/PDADMAC ratio R in two different solvents (EPR frequency: 9.79 GHz). Parameter uncertainties are within marker size.
a) In 70 vol.-% NMPA/30 vol.-% water (based on data from Ref. 5).
b) In 66 wt.-%.glycerol/34 wt.-% water. Diamonds/solid lines: low-field manifolds ($k = 1$); circles/dashed lines: center-field manifolds ($k = 2$); triangles/dash-dot lines: high-field manifolds ($k = 3$). Values for the center-field line of pure FS are also given, lines are meant as guide to the eye.
c) Stretch factors x as characterization of relaxation time distribution (see Equation (1)). Straight solid lines are linear fits including data points up to $R = 0.067$ (marked by vertical dotted line; extrapolation of the linear fits: dash-dot lines). Linear fits: in NMPA/water: $x = 0.889 - 2.019R$; in glycerol/water: $x = 0.847 - 2.353R$.

The averaged relaxation times $\langle T_{2,k}\rangle$ of the fits to the EPR spectral line manifolds in glycerol/water are significantly lower than their respective values in NMPA/water. Plots of the $\langle T_{2,k}\rangle$ for both solvent systems show that the relaxation times – unlike the stretch factors –

© 2004 WILEY-VCH Verlag GmbH & KGaA, Weinheim

do not show the same trends. Firstly, in glycerol/water, the values for $R = 0.125$ deviate from the trends of the relaxation times in the range $R = 0 \rightarrow 0.067$. Secondly, in this range the low-field line relaxation time is almost unaltered, whereas the center-field relaxation times decrease and the high-field relaxation times increase. In the same R-range, relaxation times of all three lines in NMPA/water steadily decrease and reach a plateau for $R > 0.067$.

Data for FS/PDADMAC in glycerol/water at $R = 0.125$ clearly deviate from the described trends. Using Equation (1), one could not fully fit the FS EPR spectrum at $R=0.125$ in glycerol/water (spectrum not shown) or at even higher R. A certain amount of the spectral intensity in these spectra is distributed over a large spectral width and is not accounted for in the fit. Thus, the extracted values represent relaxation-, distribution- and dynamic data of only a fraction ($R = 0.25$: 50%, $R = 0.125$: 85%) spin probes and cannot be compared to the data in the R-range up to 0.067.

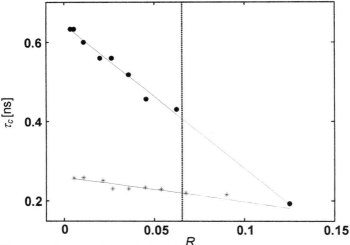

Figure 3. Rotational correlation times of FS spin probe calculated from the linewidth differences of high-field and low-field line manifolds (see text) as function of FS/PDADMAC ratio R in different solvents. Filled circles: in glycerol/water (τ_c(only FS) = 0.025 ns); asterisks: in NMPA/water (τ_c(only FS): 0.011 ns). Solid lines are linear fits including data points up to $R = 0.067$ (marked by vertical dotted line; extrapolation of linear fits: dash-dot lines). Linear fits: in glycerol/water: τ_c=0.644 ns – 3.630 ns*R; in NMPA/water: τ_c=0.258 ns – 0.616 ns*R.

Despite the obvious differences, FS/PDADMAC spectra in both solvents have a certain feature in common, namely that the differences between the average relaxation times of the high-field line ($k = 3$) and the low-field line ($k = 1$) become smaller with larger R-values. This corresponds to on average faster rotational motion of the counterion probe with decreasing

© 2004 WILEY-VCH Verlag GmbH & KGaA, Weinheim

polyelectrolyte content of the solution. In Figure 3 rotational correlation times $\langle \tau_c \rangle$ for FS in both solvent systems are plotted as a function of R. The decrease in $\langle \tau_c \rangle$ with increasing R is much larger in glycerol/water ($\langle \tau_c \rangle$ decreases by ~66%) as compared to NMPA/water ($\langle \tau_c \rangle$ decreases by ~20%). The slowdown of rotational diffusion with increasing polyelectrolyte content is linear in R in both solvents.

TAM and PDADMAC in Glycerol/Water and Water

CW EPR spectra of the spin probe TAM consist of only one single line because of lack of hyperfine interaction of the electron spin with magnetic nuclei.[17] Typical spectra for the TAM probe with different concentrations of PDADMAC in glycerol/water are presented in Figure 4a. Clearly broadening in the EPR spectra increases with increasing R. Averaged relaxation times $<T_2>$ and stretch factors x that were gained from fitting spectra to Equarion (1) are dis-

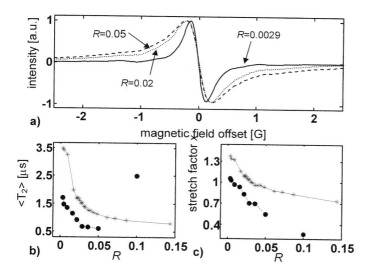

Figure 4. Representative EPR spectra, averaged relaxation times $<T_2>$, and stretch factors x (see Equation (1) and (2)) for spin probe TAM as function of TAM/PDADMAC ratio R in two different solvents (EPR frequency: 9.79 GHz). Parameter uncertainties are within marker size.
a) Three EPR spectra of TAM in glycerol/water at different R values; solid line: $R = 0.0029$; dotted line: $R = 0.02$; dashed line: $R = 0.05$; spectra are normalized to equal amplitude.
b) Averaged relaxation times $<T_2>$; filled circles: in glycerol/water ($<T_2>$(only FS) = 3.49μs); asterisk: in water ($<T_2>$(only FS) = 4.65μs).
c) Stretch factors x as characterization of relaxation time distribution (see Equation (1); filled circles: in glycerol/water (x(only FS) = 1.31); asterisks: in water (x(only FS) = 1.48). Lines are meant as guide to the eye.

© 2004 WILEY-VCH Verlag GmbH & KGaA, Weinheim

played in Figure 4b,c. The trends not only for $<T_2>$, but also for the stretch factors x are similar in both solvents, water and glycerol/water, in the range $0 < R \leq 0.05$. The only difference is that both quantities are shifted to lower values for glycerol/water.

As was the case with FS, the value for $<T_2>$ in glycerol/water at the smallest polyelectrolyte concentration (here $R=0.1$) deviates from the trend at lower concentration. This can again be attributed to our model not being able to completely fit the experimental data in glycerol/water at such R-values.

Discussion

General consequences of the observed spectral changes for a model of counterion distribution and dynamics

As the EPR transition frequency of the TAM spin probe at X-band frequencies (~9.7 GHz) does not significantly depend on orientation, the strong line broadening on addition of polyelectrolyte can only be due to a change in the environment of the probe, but not due to changes in dynamics. The observed strong changes with maximum broadening at low polyelectrolyte concentration are easily rationalized by enrichment of this counterion probe close to the polyelectrolyte chain, resulting in more frequent collisions and thus in Heisenberg exchange broadening. Any alternative explanations such as for instance effects of different local oxygen concentration could hardly explain changes of the observed magnitude. The deviation from a Lorentzian lineshape then proves a heterogeneous distribution of the TAM probes, i.e. a concentration gradient.

The same effect is apparent in the spectra of the FS spin probe, but in this case the spectrum is in addition sensitive to rotational diffusion. The presence of polyelectrolyte causes a moderate slowdown of this motion, with apparent rotational correlation times increasing with increasing polyelectrolyte concentration and never exceeding 1 ns.

These findings exclude the most simple model of counterion condensation in which one fraction of the counterions forms long-lived (lifetime $\geq$ 1 ns), static *contact ion pairs* with ionic groups of the polyelectrolyte while the remainder freely diffuses in solution. In such a case, the condensed counterions would contribute a spectral component analogous to spin labels attached to a polymer chain (restricted sidegroup motion, no isotropic averaging on the EPR timescale), while the freely diffusing probes would contribute a narrow component with the same τ_c as observed in the absence of polyelectrolyte. Such bimodal spectra are clearly not what we observe. In contrast, isotropic averaging (i.e. almost unhindered rotation about

© 2004 WILEY-VCH Verlag GmbH & KGaA, Weinheim

all three molecular axes) for the vast majority of the spin-carrying counterions on a sub-nanosecond timescale is evidence for only temporary electrostatic attachment to the chains in solution. This is a finding already indicated in our previous work,[7] but reinforced now by the observation that even at higher solvent viscosity and thus both slower polyelectrolyte chain dynamics and slower translational diffusion of the counterions there is no indication for bimodal spectra.

Note that contact ion pairs consisting of two ions with different charge have been extensively discussed before in the context of solutions of low molecular weight electrolytes (see, e.g., Ref. [18]) and that for ion pairs formed after covalent bond breaking the dynamics of dissociation and reencounter have been studied by picosecond absorption spectroscopy.[19] Results of the latter study show that the lifetime of ion pairs consisting of a diphenylmethyl cation and a chloride anion in acetonitrile is on the order of 150 ps. A model of dynamic electrostatic attachment in which at any given time there exists a fraction of contact ion pairs with such a short (sub-nanosecond) lifetime could be reconciled both with Manning theory,[8,21] and with our present results. From the experiments and analysis undertaken in this paper it is not possible to determine the molecular structure of the short lived contact ion pairs between FS spin probes and PDADMAC monomeric units. Results of experiments on shock-frozen solutions, which allow for such characterization, will be published elsewhere.[20]

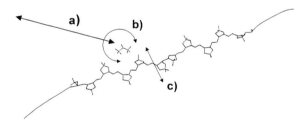

Figure 5. Sketch of the domain of an FS spin probe electrostatically attached to a PDADMAC chain. This is meant as a snapshot of the state when the spin probe is very close to the chain. Three dynamic processes are important:
a) attachment – detachment dynamics;
b) rotational dynamics of FS spin probe;
c) motion of PDADMAC polyelectrolyte chains.
A fully stretched polymer conformation is not implied.

The CW EPR spectra discussed here depend on local concentration and rotational correlation time of the spin-carrying counterions. A qualitative or semi-quantitative discussion of their

© 2004 WILEY-VCH Verlag GmbH & KGaA, Weinheim

dependence on R should thus consider the counterion concentration gradient in the vicinity of the chains and the three dynamic processes indicated in Figure 5:

    a) attachment/detachment of counterion probes due to translational diffusion,

    b) rotational diffusion of freely diffusing spin probes, and

    c) reorientation of attached counterion probes due to polyelectrolyte chain dynamics.

While full quantification of such a model is beyond the scope of this work, we shall try to identify which of the processes dominate spectral lineshapes in which regimes of solvent viscosity and polyelectrolyte concentration.

Data Interpretation: FS and PDADMAC

It has been investigated whether the influence of solvent composition on polyelectrolyte chain conformation induces changes in the dynamic electrostatic attachment of probe ions to the polyelectrolyte.[7] In solvent mixtures such as 70 vol.-% NMPA/30 vol.-% water, which has a relative permittivity higher than pure water ($\varepsilon_r \approx 140$ at 293 K, pure water: $\varepsilon_r \approx 80$ at 293 K), or 50 vol.-% ethanol/50 vol.-% water, which has a relative permittivity lower than pure water ($\varepsilon_r \approx 50$ at 293 K), analysis of the EPR lineshapes remarkably gave very similar results, while the results in pure water deviated strongly from those of the two aforementioned systems. This indicated that relative permittivity is not the most important solvent property in our system.[7] Consistent with our data, we concluded that only in water a polyelectrolyte chain conformational transition (as a second-order type transition) takes place from an extended conformation to a more globular conformation. We attributed the absence of such a counterion-induced chain collapse in the solvent mixtures of water and organic solvent with both higher and lower permittivity to screening of intramolecular hydrophobic interactions between repeat units of the polyelectrolyte chains by the organic solvent, which preferentially solvates the polyelectrolyte backbone.

Against this background we now compare the low-viscosity system in NMPA/water and the higher viscosity system in glycerol/water (Figure 2).

It is obvious that for NMPA/water (Figure 2a) $<T_{2,k}>$ decreases for all k with increasing R and reaches a plateau value for $R > 0.067$. This can be explained as follows: Assuming that over the long run all divalent probe ions partake in the dynamic attachment to the polyelectrolyte,[7] the average amount of charged PDADMAC repeat units per probe ions decreases with increasing R. Thus both the average local concentration and the heterogeneity of environments (i.e. the concentration gradient from the polyelectrolyte chains towards the free solution) increase with increasing R. The increase in local concentration is accompanied

© 2004 WILEY-VCH Verlag GmbH & KGaA, Weinheim

by spin-exchange coupling starting to dominate line broadening.[16] Increased heterogeneity is exhibited in an increasing deviation of the stretch factor x from unity, the distribution of relaxation times becomes broader, the stretch factor deviates more strongly from unity (asterisks in Figure 2c).

In contrast, the acceleration of rotational diffusion with increasing R (asterisks in Figure 3) causes a narrowing of the lines which is most pronounced for the high-field line and least pronounced for the center line. Indeed, the decrease of $\langle \tau_c \rangle$ is expected for a dynamic equilibrium between the slowly reorienting attached state and the fast reorienting detached state, as larger R and thus lower polyelectrolyte concentration correspond to a smaller fraction of time spent in the attached state. The monotonic decrease of all the $<T_{2,k}>$ with increasing R shows that in NMPA/water Heisenberg exchange broadening dominates over dynamic broadening for all lines.

The situation is slightly different for the system in glycerol/water (Figure 2b). Trends similar to those in NMPA/water are observed only for the stretch factor x and the averaged relaxation time $<T_{2,2}>$ of the center-field line, which is the line least affected by changes in the rotational correlation time.[16] The decrease in relaxation time in the range $0 < R \leq 0.067$ for this line still reflects increased spin-exchange coupling and local concentration of spin probes, the same behavior as found for NMPA/water as solvent. The plateau observed at very low R values with the decrease starting only after the third data point can be attributed to the fact that at our lowest R values polyelectrolyte concentration is so high that we are in an *overlap regime* with respect to the polyelectrolyte chains. Local counterion probe concentration approaches the bulk concentration in this regime.

The trend for the stretch factor x (filled circles in Figure 2c) also indicates similar behavior in glycerol/water as in NMPA/water. Center-field relaxation times $<T_{2,2}>$ and stretch factors x thus indicate that PDADMAC chain conformation in glycerol/water is similar to that in NMPA/water and that no conformational change takes place in the studied range of concentrations.

Exchange broadening is expected to depend on viscosity, as the frequency of molecular collisions depends on the diffusion coefficient. For free diffusion and assuming that the Stoke-Einstein relation holds, the exchange frequency and thus also the linewidth is proportional to the inverse viscosity, so that $<T_{2,2}>$ should be proportional to the viscosity.[13] However, what we observe is a shorter $<T_{2,2}>$ (i.e. larger linewidth) in the glycerol/water mixture (higher viscosity) than in the NMPA/water mixture (lower viscosity) as well as a larger deviation of the stretch factor x from unity. In particular the latter finding indicates that

© 2004 WILEY-VCH Verlag GmbH & KGaA, Weinheim

slower diffusion in the more viscous solvent causes build-up of a larger concentration gradient, so that local counterion concentration in the immediate vicinity of the polyelectrolyte chain would also be larger.[22] The collision frequency is proportional to local concentration, and the concentration increase may thus compensate or even overcompensate the decrease due to slowing down of translational diffusion. Note however, that part of the stronger broadening of the center line is due to slower rotational diffusion.[15, 23, 24]

The trend for the low-field ($<T_{2,1}>$) and high-field $<T_{2,3}>$ averaged relaxation times in glycerol/water clearly differs from the one in NMPA/water. In the glycerol/water mixture these linewidths are thus dominated by slow rotational diffusion. Consequently this solvent mixture is better suited for characterizing dynamic attachment, while the NMPA/water (or ethanol/water) mixture is better suited for characterizing the concentration gradient.

The larger sensitivity of the glycerol/water mixture to dynamics rather than concentration gradients is manifest in the plot of $\langle \tau_c \rangle$ (Figure 3). In the R-range from 0 to 0.067 we obtain good fits for a linear dependence of $\langle \tau_c \rangle$ on R for both solvent systems but the slope of the decay for glycerol/water is sixfold as compared to the one for NMPA/water. It is also apparent that the slowdown of rotational diffusion in glycerol/water mixtures with respect to NMPA/water is much more dramatic at small R values than in the absence of the polyelectrolyte (slowdown by only a factor of 2.3) or in the presence of only a small concentration of polyelectrolyte. Therefore, this slowdown cannot be attributed to the viscosity effect on rotational diffusion that is described by the Stokes-Einstein relation or a modified version of this theory.[23, 24] Indeed, a stronger dependence of $\langle \tau_c \rangle$ on solvent viscosity is expected for dynamically attached counterions than for freely diffusing counterions for two reasons. First, slower translational diffusion may increase the lifetime of the attached state. Second, influence of the viscosity is much more dramatic for the polyelectrolyte molecule with its large radius of gyration than for a small probe. Residual motion in the attached state is thus dramatically reduced.

TAM and PDADMAC

The TAM spin probe is trivalent, much larger, and tristar-shaped with a large separation of the ionic groups (~1.1 nm) , so that direct comparison between data with TAM and FS as spin probe is not possible. On the other hand, linewidths are determined exclusively by exchange broadening for this probe, which increases reliability of conclusions on concentration gradients.

© 2004 WILEY-VCH Verlag GmbH & KGaA, Weinheim

In fact, the results for this probe (Figure 4b,c) confirm our conclusions on exchange broadening for the FS probe. The trends for averaged relaxation times and stretch factors are similar in both solvents, water and glycerol/water. As in the case of FS, the decrease of $<T_2>$ and the larger deviation of the stretch factor x from unity point to build-up of a larger concentration gradient in the higher-viscosity solvent compared to the lower-viscosity solvent. However, both effects are more dramatic for the larger TAM probe and slowdown of rotational diffusion cannot explain the strong decrease in $<T_2>$ in this case. For TAM, the data thus prove rather than only indicate that the effect of an increase in local concentration in the vicinity of the polyelectrolyte on the exchange frequency overcompensates the effect of a decrease in the translational diffusion rate.

There is a sharp transition of $<T_2>$ from a steep decrease to a plateau when increasing R above ≈ 0.03. This may correspond to a transition from a strong *overlap regime* that is dominated by TAM molecules building physical crosslinks between PDADMAC chains ($R < 0.03$) to a regime in which TAM molecules interact with single polyelectrolyte chains and force the chains to form globular structures: each TAM by effectively screening three charges of the polymer and bending the chain in one domain, or by crosslinking remote domains of one single chain. This may lead to an approach to a maximum local concentration of TAM counterions that can accommodated by the chain. A further decrease of the polyelectrolyte concentration might then lead to the appearance of a fraction of counterions that interacts more weakly with only the surface of the crosslinked PDADMAC globules and thus gives rise to a spectral component with a smaller linewidth as seen at $R = 0.1$.

Conclusions

The viscosity dependence of the EPR spectra of counterion probes in polyelectrolyte solutions of varying concentration confirms our model of counterion condensation as a fast dynamic process with lifetimes of the contact ion pairs being shorter than 1 ns. While EPR lineshapes of nitroxide counterion probes in low-viscosity solvents and solvent mixtures are dominated by exchange broadening due to increased local concentration in the vicinity of the polyelectrolyte chain, they are more strongly influenced by slowdown of rotational diffusion in the high-viscosity solvent glycerol/water. The dramatic effect of an increase in polyelectrolyte concentration on the rotational correlation time can be explained by a longer lifetime of the attached state due to slower translational diffusion combined with much slower dynamics in the attached state due to the slowdown of polymer chain dynamics. Analysis of the viscosity dependence of the exchange broadening for both the FS and TAM spin probes

© 2004 WILEY-VCH Verlag GmbH & KGaA, Weinheim

indicates that slowdown of translational diffusion in more viscous solvents causes the build-up of a larger counterion concentration gradient in the vicinity of the chain.

Acknowledgments. We thank Nycomed Innovations AB, Sweden, for generously supplying us with the TAM radical used in this study and Christian Bauer for technical support. Financial support from the priority program "High-Field EPR in Biology, Chemistry and Physics" (SPP 1051) by the Deutsche Forschungsgemeinschaft (DFG) is gratefully acknowledged.

[1] Y. Nishiyama, M. Satoh., *Polymer* **2001**, *42*, 3919.
J. Kötz, S. Kosmella, T. Beitz, *Prog. Polym. Sci.* **2001**, *26*, 1199.
[2] D. Voet and J. G. Voet, *Biochemistry* (2nd edition), chapter 29, J. Wiley & Sons, Inc., New York (1995) and references therein
[3] Y. Kantor, M. Kardar, *Phys. Rev. E* **1995**, *51*, 1299.
A. V. Dobrynin, M. Rubinstein, S. P. Obukhov, *Macromolecules* **1996**, *29*, 2974.
H. J. Limbach, C. Holm, K. Kremer, *Europhys. Lett.* **2002**, 60, 566.
[4] N. M. Atherton, *"Principles of Electron Spin Resonance"*, Ellis Horwood Limited, New York 1993.
[5] L. Columbus, W. L. Hubbell, *Trends Biochem. Sci.* **2002**, *27*, 288.
H. J. Steinhoff, A. Savitsky, C. Wegener, M. Plato, K. Möbius, *Biochim. Biophys. Acta* **2000**, *1457*, 253.
T. M. Okonogi, S. C. Alley, E. A. Harwood, P. B. Hopkins, B. H. Robinson, *Proc. Nat. Ac. Sci.* **2002**, *99*, 4156.
[6] M. C. Café, I. D. Robb, *Polymer* **1979**, *20*, 513.
H. Q. Xue, P. Bhowmik, S. Schlick, *Macromolecules* **1993**, *26*, 3340.
A. M. Wasserman, V. A. Kasaikin, V. P. Timofeev, *Spectrochimica Acta, Part A* **1998**, *54*, 2295.
S. E. Cramer, G. Jeschke, H. W. Spiess, *Macromol. Chem. Phys.* **2002**, *203*, 182 and 192.
[7] D. Hinderberger, G. Jeschke, H. W. Spiess, *Macromolecules* **2002**, *35*, 9698.
[8] G. S. Manning, *Acc. Chem. Res.* **1979**, 12, 443.
[9] G. W. Scherer, *"Relaxation in glass and composites"*, chapter 4, Krieger, Malabar Florida 1992.
[10] J. S. Hyde, A. Jesmanowicz, J. J. Ratke, W. E. Antholine, *J. Mag. Res.* **1992**, *96*, 1.
[11] C. P. Lindsey G. D. Patterson, *J. Chem. Phys.* **1980**, *73*, 3348.
[12] A. Schweiger, G. Jeschke, *"Principles of Pulse Electron Paramagnetic Resonance"*, Oxford University Press, Oxford 2001, p. 97.
[13] Y. N. Molin, K. M. Salikhov, K. I. Zamaraev, *"Spin Exchange: Principles and Applications in Chemistry and Biology"*, Springer-Verlag, Berlin 1980.
[14] K. M. Salikhov, A. G. Semenov, Y. D. Tsvetkov, *"Electron Spin Echo and Its Applications"*, chapter 4, Nauka, Novosibirsk 1976, pp 195-235.
[15] J. S. Hwang, R. P. Mason, L. P. Hwang, J. H. Freed, *J. Phys. Chem.* **1975**, *79*, 489.
S. A. Goldman, G. V. Bruno, C. F. Polnaszek, J. H. Freed, *J. Chem. Phys.* **1972**, *56*, 716.
[16] B. H. Robinson, C. Mailer, A. W. Reese, *J. Mag. Res.* **1999**, *138*, 199 and 210.
[17] J. H. Ardenkjaer-Larsen, I. Laursen, I. Leunbach, G. Ehnholm, L. G. Wistrand, J. S. Petersson, K. Golman, *J. Mag. Res.* **1998**, *133*, 1.
Lu. Yong, J. Harbridge, R. W. Quine, G. A. Rinard, S. S. Eaton, G. R. Eaton, C. Mailer, E. Barth, H. J. Halpern, *J. Mag. Res.* **2001**, *152*, 156.
[18] R. Behrends, P. Miecznik, U. Kaatze, *J. Phys. Chem. A* **2002**, *106*, 6039.
[19] K. S. Peters, B. Li, *J. Phys. Chem.* **1994**, *98*, 401.
[20] D. Hinderberger, H. W. Spiess, G. Jeschke, *submitted.*
[21] G. S. Manning, *J. Chem. Phys.* **1969**, *51*, 924
[22] Such an effect may not be expected in the Manning picture of a fully stretched static chain with constant electrostatic potential, but might be a consequence of motion of both the polyelectrolyte chain and the counterions: the electrostatic potential of such a chain does depend on its segmental motion.
[23] R. E. D. McClung, D. Kivelson, *J. Chem. Phys.* **1968**, *49*, 3380.
[24] This effect is actually not as strong as Stokes-Einstein theory predicts. Hwang et al. (Ref. [10]) have shown that a nitroxide spin probe rotated more than seven times faster in pure glycerol than expected from bulk viscosity. In our case the slowest rotational correlation times in glycerol/water are only three times as large as in NMPA/water. McClung and Kivelson (Ref. [16]) introduce a relation for the rotational correlation time which includes an empirical *slip parameter*. This parameter κ ($0 \leq \kappa \leq 1$) accounts for rotational motion faster than expected from bulk viscosity: $\tau_c = \kappa^* \tau_c$ (Stokes-Einstein).

© 2004 WILEY-VCH Verlag GmbH & KGaA, Weinheim

Macromol. Symp. **2004**, *211*, 87-92

Counterion Mobility and Effective Charge of Polyelectrolytes in Solution

Ute Böhme, Ulrich Scheler*

Institute for Polymer Research Dresden e. V., Hohe Strasse 6,
D-01069 Dresden, Germany

Summary: Polyelectrolytes are macromolecules containing dissociable or charged groups. The charge, that is effectively accessible, is determined by counterion condensation, which is strongly influenced by the ionic strength of the solution under study. In general a rapid exchange between free and condensed counterions is expected. In the present study diffusion and electrophoretic mobility of poly(diallyldimethylammoniumchloride) and perfluorinated succinic acid have been monitored simultaneously. Condensation of the perfluorinated succinic acid to the macroion shows in the electrophoretic mobility of succinic acid monitored by pulsed field gradient NMR. In the concentration dependence of the averaged diffusion coefficient and electrophoretic mobility an exchange fast on the time scale of the NMR experiment is manifested.

Keywords: counterion condensation; effective charge; electrophoretic mobility; polyelectrolytes; pulsed field gradient NMR

Introduction

Polyelectrolytes are polymers containing charged groups, therefore their solution properties are strongly determined by the electrostatic interaction between these charges. However, the effective charge of strong polyelectrolytes is determined by counterion condensation rather than the nominal charge. If the charge-charge interaction would exceed the thermal energy, counterions condense onto the macromolecule, generating the force balance.[1] Condensation of counterions is enhanced upon the addition of salt to a solution of polyelectrolytes.[2]

Pulsed field gradient NMR (PFG NMR)[3, 6] is applied to probe molecular displacements of both the macromolecules and counterions in solution.[7] From the diffusion experiment the friction coefficient for the velocity-proportional friction is derived according to the Einstein equation. The electrophoresis NMR experiment yields the electrophoretic mobility.[8, 9] From both the effective charge of the macromolecules can be determined assuming, that the molecules are free draining, which is justified because of the fractal dimension of the polymers well below 2.[10] In the following a qualitative picture is discussed where the selectivity of the NMR is utilized with a special emphasis on the counterions.

© 2004 WILEY-VCH Verlag GmbH & KGaA, Weinheim

DOI: 10.1002/masy.200450706

Experimental

The experiments have been performed on a Bruker Avance 300 NMR spectrometer, operating at Larmor frequencies of 300 MHz and 282 MHz for protons and fluorine respectively. The spectrometer is equipped with a micro imaging accessory providing magnetic field gradients up to 1 T/m. The electrophoresis NMR system is in-house built and is described elsewhere in more detail.[7] The new system, permitting the investigation of small molecules simultaneously, consists of a U-shaped tube with the NMR coil on one side of the U. Any possible NMR signal from the other side is blocked by an concentric rf shield around the rf coil. The electric field is generated by a pair of platinum electrodes on top of the U. Placing the electrodes on the top permits any gas produced by a possible electrode reaction to escape without disturbing the experiment. The electrodes are connected to a custom made DC amplifier which is driven by one of the gradient channels of the spectrometer to synchronize the DC field with the NMR experiment.

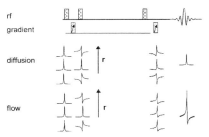

Figure 1. Schematic illustration of the pulsed field gradient stimulated echo NMR experiment. On the bottom the signal phase for three locations is shown schematically indicating the resulting signal attenuation in the case of incoherent motion and phase modulation in the case of coherent motion respectively.

A stimulated echo pulsed field gradient NMR experiment has been used to measure diffusion. From the echo attenuation the diffusion coefficient is derived using the Stejskal-Tanner equation.[6] In the electrophoresis NMR experiment the duration of the gradient pulses is kept constant and the strength of the electric field as the driving force for the electrophoretic motion is incremented.[10, 7]

For ^{1}H-diffusion measurements z-field gradients varied in 32 or 64 steps from 0.12 T/m to 0.98 T/m with 6 ms gradient pulse and 40 ms diffusion delay were applied. For ^{19}F-diffusion measurements the z-field gradients were incremented from 0 T/m to 0.98 T/m for all solutions containing PDADMAC and from 0 T/m to 0.5 T/m for samples without it, respectively. The gradient pulse duration δ was 4 ms and the diffusion delay Δ 20 ms. All electrophoresis

© 2004 WILEY-VCH Verlag GmbH & KGaA, Weinheim

experiments were performed incrementing the electric field in 16 steps from   -120 V/cm to +120 V/cm or conversely. ^1H was measured with a gradient pulse duration δ of 6 ms, a diffusion delay Δ of 28 ms and a constant magnetic field gradient of 0.8 or 0.9 T/m. For ^{19}F-electrophoresis experiments the same time constants as for diffusion have been used. The applied z-field gradient varied from 0.3 to 0.9 T/m, depending on the diffusion behaviour of the sample.

For each value of the gradient strength or the electric field respectively an NMR spectrum has been detected. Thus components of the solution are identified from the spectrum. In the proton NMR spectrum signals from the polymer and a residual signal from H_2O in D_2O have been observed. In the fluorine spectrum only perfluoro succinic acid has been observed.

Poly(diallyldimethylammoniumchloride) (PDADMAC, Mw = 240kg/mol, Polyscience) and the perfluorinated acid (perfluorinated succinic acid PFSA, Aldrich) have been used as received without further purification. A 5.2 mmol/l (monomer) solution of the polyelectrolyte in D_2O was used for all experiments. Starting from this solution, small quantities of a high concentrated perfluorinated succinic acid solution were added, so that the volume expansion could be neglected. The concentrations of the dibasic acid PFSA adjusted were 1.3, 1.9, 2.6 and 3.9 mmol/l, resulting in 2.6, 3.9, 5.2 and 7.8 mmol/l equivalent charges, respectively.

Results

Typical NMR spectra of poly(diallyldimethylammoniumchloride) and perfluorinated succinic acid are depicted in Figure 2.

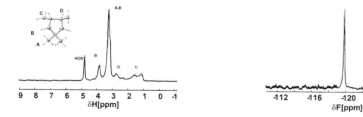

Figure 2. One-dimensional NMR spectra of poly(diallyldimethylammoniumchloride) (left) and perfluorinated succinic acid (right).

Although the resolution in these spectra is not comparable with that in high-resolution NMR spectra, it is sufficient for signal assignment. In the following the selection will be based on the observed nucleus, where in the proton spectra signals from poly(diallyldimethyl-ammoniumchloride) and in the fluorine spectra the signal from perfluorinated succinic acid are observed respectively.

© 2004 WILEY-VCH Verlag GmbH & KGaA, Weinheim

Figure 3 depicts two-dimensional electrophoresis NMR spectra correlating chemical shift with electrophoretic mobility. According to Ref. [10], a model-free data evaluation based on a two-dimensional Fourier transform permits the assignment of the electrophoretic mobility and its sign.

The electrophoretic mobility of the polycation by definition is positive. The mobility of the fluorinated acid in the free solution is negative. If the acid is added to the solution of the polycation in a concentration well below theoretical charge compensation, the effective mobility, that is measured for the acid, is positive as well. This reversal of the effective electrophoretic mobility is a direct proof of the condensation of the perfluorinated succinic acid as a counterion onto the polycation.

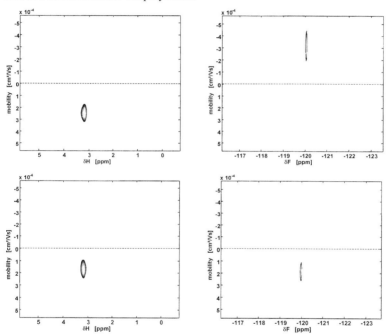

Figure 3. Two-dimensional electrophoresis NMR spectra of 5.2 mmol/l (monomer) PDADMAC (left) and 2.6 mmol/l perfluorinated succinic acid (right) for the free components (top) and a mixture of 5.2 mmol/l (monomer) PDADMAC with 1.3 mmol/l perfluorinated succinic acid, that represents compensation of half the charges of the polycation by the acid (bottom).

When the concentration of the acid is increased a decreasing effective electrophoretic mobility is observed. The term effective is emphasized here, because in the NMR experiment the average between the condensed and free acid is observed, when the exchange between the

© 2004 WILEY-VCH Verlag GmbH & KGaA, Weinheim

two populations is rapid on the time scale of the experiment, which here is in the order of 10 ms. Although increasing the total concentration of the fluorinated acid by more than a factor of four will lead to enhanced counterion condensation[2] most of the additional acid will increase the population of the free acid in the solution. Therefore the effective electrophoretic mobility shifts towards the mobility of the free acid. The same is true for the effective, exchange averaged diffusion coefficient as depicted in Figure 4. The diffusion coefficient and the electrophoretic mobility of poly(diallyldimethylammoniumchloride) remain nearly unchanged.

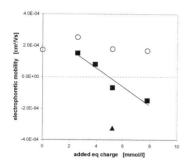 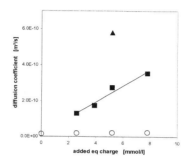

Figure 4. Effective electrophoretic mobility (left) and diffusion coefficient (right) of (■) perfluorinated succinic acid, (▲) perfluorinated succinic acid without PDADMAC and (○) PDADMAC, measured as a function of the charge added.

Conclusions

It has been demonstrated, that the electrophoretic mobility of both, poly(diallyldimethylammoniumchloride) and a perfluorinated acid are determined simultaneously in an electrophoresis NMR experiment. Both compounds in pure solution exhibit electrophoretic mobility of opposite signs. If the perfluorinated acid is added to the polymer in a concentration well below charge compensation, an effective electrophoretic mobility close to that of the polymer is detected, indicating that the perfluoro succinic acid replaced the originally condensed counterions on the polymer. Increasing the concentration results in a shift of both the electrophoretic mobility and the diffusion coefficient towards the value of the free acid. The fact, that a single component in both is detected at all concentrations proofs, that there is fast exchange between free and condensed acid on the time scale of the pulsed field gradient NMR experiment.

© 2004 WILEY-VCH Verlag GmbH & KGaA, Weinheim

92

Acknowledgement

This work has been supported by the Deutsche Forschungsgemeinschaft (DFG) under grant SCHE 524/2-2.

[1]   G. S. Manning, in *"Polyelectrolytes"*, E. Sélégny, M. Mandel, U. P. Strauss, Eds., D. Reidel Publishing Company, Dordrecht-Holland 1974.
[2]   U. Böhme, U. Scheler, *Colloids and Surfaces A* **2003**, *222*, 35.
[3]   P. T. Callaghan, *"Principles of Nuclear Magnetic Resonance Microscopy"*, Oxford University Press, Oxford 1991.
[4]   R. Kimmich, *"NMR Tomography, Diffusometry, Relaxometry"*, Springer, Berlin, Heidelberg 1997.
[5]   J. Kärger, in *"Diffusion in Condensed Matter"*, Kärger, Heitjans, Haberland Eds., Vieweg, Braunschweig 1998.
[6]   E. O. Stejskal, J. E. Tanner, *J. Chem. Phys.* **1965**, *42*, 288.
[7]   S. Wong, U. Scheler, *Colloids and Surfaces A* **2001**, *195*, 253.
[8]   M. Holz, *Chem. Soc. Rev.* **1994**, *23*, 165.
[9]   C. S. Johnson, in *"Encyclopedia of Nuclear Magnetic Resonance"*, D.M. Grant, R.K. Harris Eds., John Wiley & Sons, Chichester 1996, and references therein.
[10]   M. Ullner, in *"Handbook of Polyelectrolytes and Their Applications"*, S. K. Tripathy, J. Kumar, H. S. Nalwa Eds., American Scientific Publishers 2002, Vol. 3, 271.
[11]   U. Scheler, in *"Handbook of Polyelectrolytes and Their Applications"*, S .K. Tripathy, J. Kuma, H. S. Nalwa Eds., American Scientific Publishers 2002.

© 2004 WILEY-VCH Verlag GmbH & KGaA, Weinheim

Poly(Styrene Sulfonate) Self-Organization: Electrostatic and Secondary Interactions

H. Ahrens,[1] K. Büscher,[1,2] D. Eck,[2] S. Förster,[3] C. Luap,[4,5] G. Papastavrou,[1] J. Schmitt,[2] R. Steitz,[5,6] C. A. Helm [1,2]

[1] Institut für Physik, Universität Greifswald, D-17489 Greifswald, Germany

[2] Institut für Physikalische Chemie, Universität Mainz, D-55099 Mainz, Germany

[3] Institut für Physikalische Chemie, Universität Hamburg, D-20146 Hamburg, Germany

[4] Stranski-Laboratorium, TU Berlin, Straße des 17. Juni 112, D-10623 Berlin, Germany

[5] BENSC, Hahn-Meitner Institut, Glienicker Straße 100, D-14109 Berlin, Germany

[6] MPI of Colloids and Interfaces, D-14424 Potsdam, P.O. Box 5607, Germany

Summary: We investigate the self-organization of PSS in brushes and polyelectrolyte multilayers with X-ray, neutron and optical reflectivity. The electrostatic force dominates brush phases and adsorption behavior, additionally we find evidence of a strong hydrophobic force: (ι) within amphiphilic diblock copolymer monolayers, a PSS monolayer adsorbs flatly to the hydrophobic block, (ιι) on temperature increase (and with screened electrostatic forces), more PSS is adsorbed onto oppositely charged surfaces, and (ιιι) a polyelectrolyte multilayers shrinks when heated at 100% r.h. The latter two effects are consistent with the well-known increase of the hydrophobic force on heating: The increased PSS surface coverage can be attributed to deteriorating solvent conditions. Within a polyelectrolyte multilayer, an increase of the hydrophobic force maximizes the local contact of hydrophobic polymer segments, causing a reduction of swelling and an increased mass density.

Keywords: monolayers; polyelectrolytes; self-organization

© 2004 WILEY-VCH Verlag GmbH & KGaA, Weinheim   DOI: 10.1002/masy.200450707

Introduction

Polyelectrolytes remain among the least understood materials in condensed matter science, despite their widespread presence and use. They are difficult to understand because of the entwined correlations of chain configuration and charge, coupled with the long-ranged interactions inherent to these structures. Yet, in the past decade the field of nanostructured material formation has progressed significantly. Self-assembly processes of polyelectrolytes involving electrostatic interactions can be used to build-up multilayered materials with unique properties. The basic principle is the sequential adsorption of positively and negatively charged polyelectrolytes.[1,2] Therefore, polyelectrolyte multilayers form two-dimensionally stratified layers which are growing step-by-step into the third dimension. This leads to a behaviour being dominated by internal interfaces and local interactions, and differing largely from the corresponding volume material properties.

Polyelectrolyte adsorption into an oppositely charged interface is determined by electrostatic and secondary interactions.[3,4] While the electrostatic interactions are qualitatively understood, little is known about the nature, the origin and the influence of the secondary forces.[5]

The flat conformation of polyelectrolytes adsorbed from salt solutions of low concentrations (i.e. below 0.1 Mol/L) is attributed to the large amplitude and range of the electrostatic force.[3] At high salt concentrations, both the polyelectrolyte coverage and the thickness of the adsorbed layer increase, features which are attributed to the screening of the electrostatic force.[6] Therefore, the respective strength of secondary forces such as segment/interface interaction and the polymer-solvent interaction (parameter w) start to be important. For $w=0$ (good solvent conditions) entropy dominates. For large w, polymer-solvent contact is unfavourable, the chain contracts, and, for very large w, the polymer precipitates. The transition is gradual, on increase of w, more polymer is adsorbed onto a surface.[7]

If the solvent is water, apolar or hydrophobic molecular groups of the polymer induce "bad solvent conditions".[8] An apolar molecule cannot form hydrogen bonds with water molecules, so it distorts the usual water structure, forcing the water into a rigid cage of hydrogen-bonded molecules around it. Water molecules as typical solvent molecules are normally in constant motion, and cage formation restricts the motion of a number of water molecules. This effect increases the structural organization of water, and thus decreases the entropy of the water molecules. If two hydrophobic groups aggregate, only one cage needs to be formed and fewer water molecules are confined. Therefore, hydrophobic groups are attracted towards each other,

© 2004 WILEY-VCH Verlag GmbH & KGaA, Weinheim

forming so-called hydrophobic bonds. The hydrophobic interaction increases on heating, because of the increased water mobility. A prominent example are microtubules (one of the three cytoskeletal systems involved in cell motility and the determination of cell shape), which polymerize due to hydrophobic bonds. On temperature decrease to 4°C, they depolymerize. The microtubules repolymerize at 37°C (if a suitable protein as catalysator is present).

In this paper, we will focus on poly (styrene sulfonic acid) (PSS), a strongly charged polyelectrolyte with a hydrophobic backbone, which is no water-soluble if neutral.[9] It precipitates, if the temperature is increased to 60°C, an indication of strong secondary (possibly hydrophobic) interactions. We investigate the self-organization of PSS at interfaces, and explore different effects: polyelectrolyte brush phases,[10,11] polyelectrolyte multilayer build-up as function of composition and temperature of the adsorption solution,[12] polyelectrolyte multilayers and their swelling (because of the possible application as humidity sensor).[13]

Experimental Part

For polyelectrolyte multilayer preparation, PSS (M_w=84 kDa, M_w/M_n~1.1) served as polyanion, poly (allylamine) hydrocholoride (PAH, M_w=50-65 kDa) as polycation. To form an amphiphilic diblock copolymer monolayer, Poly(ethyl ethylene) (PEE) served as the hydrophobic block, PSS as the hydrophilic block (cf. Fig. 1).

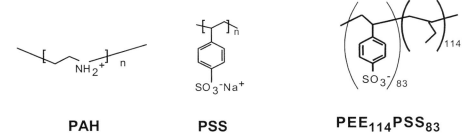

PAH   **PSS**   **PEE$_{114}$PSS$_{83}$**

Figure 1. Scheme of PAH, PSS and of PEE$_{114}$PSS$_{83}$.

Specular X-ray reflectivity experiments on solid substrates were performed with a Siemens D-500 powder diffractometer. The X-ray set-up for fluid surfaces is home-built (λ=1.54Å),[14] its angular divergence is 0.012°. The neutron set-up is experiment V6 at HMI, (wavelength λ = 4.66 Å), in the liquid scattering set-up[15] and PNR reflectometer at GKSS Research Center for solid surfaces (wavelength λ = 6.37 Å). The relative humidity (r.h.) was adjusted with salt solutions.

© 2004 WILEY-VCH Verlag GmbH & KGaA, Weinheim

For X-rays and neutrons, the specular reflection geometry is used, and the reflectivity is measured varying the angle of incidence α. The reflected intensity is measured as function of the wave vector transfer $Q_z = 4\pi \sin\alpha / \lambda$. The index of reflection n depends linearly on known material constants, and the electron density ρ or the scattering length density ρ_n, respectively: $n = 1 - r_o\rho\lambda^2 / 2\pi$ (Thompson radius $r_o = 2.82 \times 10^{-15}$ m) or $n = 1 - \rho_n\lambda^2 / 2\pi$. The reflectivity is calculated using classical Fresnel optics. To quantify the density profile, the exact matrix formalism is used.[16] In case of ambiguity, we chose the least structured profile (maximum entropy approach[17]), with which the measured reflectivity could be fitted.

Optical reflectivity experiments were performed in a UV-Vis spectrometer (Perkin-Elmer), using vertical incidence and varying wavelength. Now, the wave vector transfer corresponds to $Q_z = 4\pi / \lambda$. Depending on the environment the refractive index varies between 1.45 and 1.61. The same algorithms were used for analysis of the optical data as for X-ray and neutron reflectivity.

Atomic Force measurements (AFM) measurements were performed with a Multimode equipped with a Nanoscope IIIa (Digital Instruments; Santa Barbara, CA) in the tapping mode in air, and tips with 15 nm radius (according to the supplier, Olympus, Germany).

Results and Discussion

1. The inner interface of a PEE-PSS/water diblock copolymer monolayer

We investigated monolayers of $PEE_{114}PSS_{83}$ at the air/water interface, to characterize polyelectrolyte brushes.[10,11] Hydrophobic PEE is a suitable anchor for PSS, since it is fluid at room temperature, therefore, the PEE-PSS joints attain a new equilibrium distribution after a change in the grafting density.

To characterize the polyelectrolyte brushes, the electron density profiles of the monolayers are determined with X-ray reflectivity as function of the grafting density and the salt conditions. Extreme cases are shown in Fig. 2: pure water, or 1 M CsCl solution. X-ray reflectivity measurements are taken along the isotherm. The reflectivity curves show a very clear structure. At low Q_z, a thick layer (the PSS block) causes one narrow minimum, between two narrow maxima. On monolayer compression, the position is constant on pure water, whereas it shifts dramatically on the concentrated electrolyte solution. A thinner layer (the PEE block) causes the

© 2004 WILEY-VCH Verlag GmbH & KGaA, Weinheim

minimum at high Q_z, which shifts on increase of the grafting density to lower Q_z, indicating thickening of the PEE block.

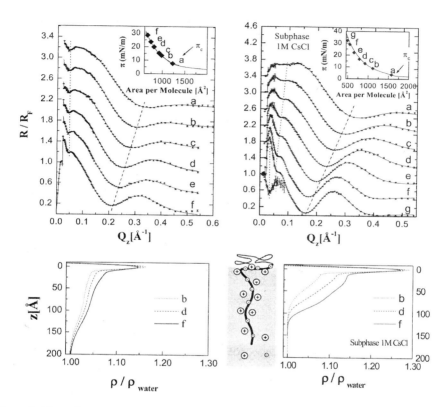

Figure 2. Normalized X-ray reflectivity from the diblock copolymer $PEE_{114}PSS_{83}$ on a subphase of pure water (top, left) or a 1M CsCl solution (top, right). The measurements were taken along the isotherm given in the inset, π_c indicates the formation of a homogeneous hydrophobic block. Below curve "g" a normalized neutron reflectivity curve taken at the same molecular area as "g" is shown. The full lines correspond to a fit of the electron (scattering length) density profiles to the data. The dashed lines are drawn through minima of equal order, as a guide to the eye. For clarity the reflectivity curves are shifted by 0.4 units. Bottom: electron density profiles of the PSS-brush at the molecular areas given in the inset. The 12Å thick layer of increased electron density is a PSS monolayer (including its counterions) which is flatly adsorbed to the PEE block.

© 2004 WILEY-VCH Verlag GmbH & KGaA, Weinheim

The neutron reflectivity data show the same periodicity as found with X-rays at the same molecular area (1 Mol/l CsCl, D_2O instead of H_2O, cf. Fig. 2). The interference pattern is shifted by π, since for neutrons the subphase exhibits a higher scattering length density than the polyelectrolyte brush, while for X-rays the situation is inverted. Yet the interference pattern can be fitted with the same segment density profile as the X-ray data (considering that the PEE-layer is not accessible for neutron scattering, due to the limited Q_z-range). Since the counterions provide most of the contrast for X-rays, whereas for neutrons the contrast is due to the polymer segments only, this is direct evidence that the counterions do indeed trace the polymer closely.

As expected, the hydrophobic PEE block shows features characteristic of a melt: independence of the subphase, a constant electron density, and a linear increase of the thickness with the grafting density. Furthermore, two polyelectrolyte brush phases can be distinguished: at low salt the osmotically swollen brush characterized by stochiometric ion incorporation and almost constant thickness; at high salt (> 0.1 M) the salted brush, which shrinks both when the molecular area or subphase salt concentration is increased, according to power laws, as theoretically predicted[18]. Unexpectedly, directly beneath the hydrophobic anchors is a ~12Å thick layer of high electron density, which is attributed to a flat monolayer of PSS-chains.[10,11] When the PEE-layer is homogeneous (i.e. above π_c) the lateral density of this monolayer is constant (1 monomer per 44 $Å^2$) and independent of the grafting density, the PSS length, the polyelectrolyte brush phase, the concentration or the type of ions. Within this monolayer, one monomer is hydrated by 5-6 water molecules, and neutralized by one counterion (in Fig. 2, the measured electron density of the adsorbed PSS monolayer depends on the counterions in the subphase, either Cs^+ with its 54 electrons, or protons with no electrons). Only when the PEE-layer is no longer homogeneous (i.e. below π_c), the lateral density of the flatly adsorbed monolayer decreases. Obviously, there is strong attraction between PSS and PEE, which can be ascribed to hydrophobic interaction.

The importance of the hydrophobic force for PSS adsorption was confirmed by other studies, such as mixed monolayers of polyelectrolytes and surfactants at the air-water interface[19], or of PSS adsorbed directly to the water surface.[6,20]

2. Build-up of polyelectrolyte multilayers at different temperatures

The formation of polyelectrolyte multilayers is generally discussed as the consequence of multiple electrostatic bonds. Yet, other attractive interactions can also contribute. As described

© 2004 WILEY-VCH Verlag GmbH & KGaA, Weinheim

above, PSS (as part of a diblock copolymer monolayer) adsorbs onto a hydrophobic surface, therefore it is likely that the hydrophobic force influences polyelectrolyte multilayer formation, too. This is studied by varying the deposition temperature under high salt conditions,[12] when both the amplitude and the range of electrostatic interactions are strongly screened,[3,21]

X-ray reflectivity curves from polyelectrolyte multilayers deposited at different temperatures are shown in Fig. 3. The thickness of the polyelectrolyte multilayer can be approximated from the average spacing Δq of the interference minima according to $d = 2\pi / \Delta q$. Two effects are obvious: on increase of build-up temperature, the interference minima shift continuously to the left, indicating film thickening. At 30°C and higher deposition temperatures, the amplitudes of the interference extrema decrease. The effect is especially obvious for higher order maxima, an effect which suggests roughening of the film/air interface. The polyelectrolyte multilayer thickness is increased by 40 % (1M KCl adsorption solution), or 20% (2M KCl), respectively.

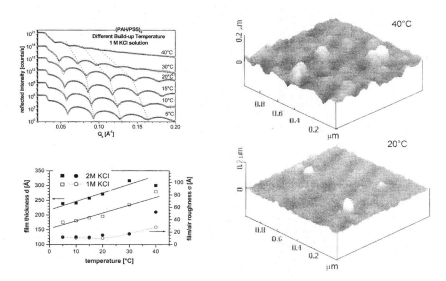

Figure 3. Top left: X-ray reflectivity curves of a 5 layer pair PSS/PAH-film, prepared from 1 M KCl solution at T=5, 10, 15, 20, 30, and 40°C, as indicated. For clarity, the X-ray reflectivity curves are shifted relative to each other, the dashed lines connecting the minima of equal order are guides to the eye. Bottom left: the film thickness (squares) and the roughness (circles) as function of the solution temperature, both for adsorption baths containing 1 M KCl (open symbols) and 2 M KCl (full symbols). Right: AFM images taken in the tapping mode of films prepared from 1 M KCl solution at 20 and 40°C, as indicated.

© 2004 WILEY-VCH Verlag GmbH & KGaA, Weinheim

To explore the lateral extension of height fluctuations on the film surface, AFM images of the multilayers are shown in Fig. 3. The lateral extension of the height fluctuations, presumably polyelectrolyte aggregates, amounts to 50 – 200 nm. With the exception of the sample prepared from 2 M KCl solution at 40°C, the roughness $\sigma = \sqrt{\left(\sum_{i=1}^{N} \left(z_i - z_m \right)^2 \right) / N}$ (z_m: average height, N: number of height measurements taken) is about a factor of two larger than the values obtained from X-ray reflectivity. Performing AFM measurements on surfaces covered with protruding polymer aggregates leads to "wrong" results with respect to the lateral aggregate dimensions, due to the convolution of the real aggregate size with the tip size.[22] Obviously, with polyelectrolyte multilayers, this convolution leads to an exaggerated roughness. The polyelectrolyte multilayers prepared at 40°C from 2 M KCl solution show a widely scattering roughness: x-ray reflectivity yields 52Å, AFM 12Å (the lowest value observed for the whole series). This scatter indicates pronounced lateral inhomogeneities, which is consistent with the severe stability problems during multilayer build-up at these conditions.

On heating a solution of macromolecules, water becomes a progressively bad solvent for hydrophobic groups[8]. For PSS we observe the analogous effect: on approaching the precipitation temperature, the polymer/solvent interaction increases. Therefore, more polyelectrolyte is adsorbed. A characteristic feature of surfaces covered with adsorbed neutral polymers is the dependence of the surface coverage on the polymer weight.[7] Yet, we do not find any molecular weight dependence,[12] indicating a still dominant role of the electrostatic forces.[3] Even though the increased layer thickness at elevated temperatures suggests an increasing importance of weak non-electrostatic bonds. Due to growing percentage of weak bonds, a polyelectrolyte molecule adsorbed at elevated temperature is more susceptible to desorption, lateral movement, etc. Indeed, we observe an increase in surface roughness.

3. Swelling of polyelectrolyte multilayers at different temperatures

To investigate the possible application of polyelectrolyte multilayers as humidity sensors, polyelectrolyte swelling was studied as function of relative humidity (r.h.), temperature and time. At 22°C, both the multilayer thickness and the water uptake (i.e. the scattering length density) increase linearly with the relative humidity, the thickness increase is about 29%,[13,23,24] corresponding to six water molecules per PAH/PSS monomer pair. With optical reflectivity,[13] the

© 2004 WILEY-VCH Verlag GmbH & KGaA, Weinheim

kinetics of the swelling and shrinking are studied. On change of the r.h., equilibrium is attained within a second. Thus the film may be described as a mesh consisting of many voids, to which water molecules adsorb or from which they desorb. The flexible cavities change their size accordingly, and cause film swelling. If one furthermore considers the linear increase of both thickness and water content on r.h. change, it makes sense to describe the swelling at constant temperature in terms of a Langmuir isotherm. This approach yields a water binding energy of $Q=33kJ/mol$.

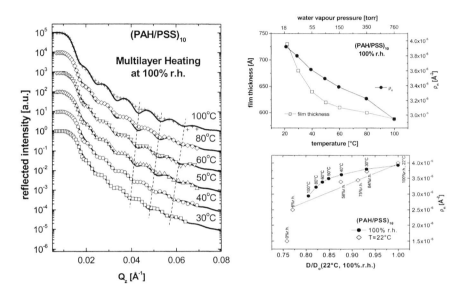

Figure 4. Left: Neutron reflectivity measurements of a 10 layer pair PSS/PAH-film (prepared at room temperature from 3 M NaCl solution) in a saturated D_2O-atmosphere at T=30, 40, 50, 80, and 100°C (bottom to top). Symbols represent data, solid lines the best fits. The reflectivity curves are shifted for the sake of clarity. The dashed lines are drawn through minima of equal order, as a guide to the eye. Top right: Film thickness (open squares) and scattering length density ρ_n (solid circles) as a function of temperature as derived from the fits shown left. Bottom right: Scattering length density as function of temperature. One film was investigated while increasing the temperature (solid circles, c.f. a,b) in a saturated D_2O atmosphere, the other one while varying the r.h. at T=22°C (open diamonds). Increasing the r.h. from 0 to 6% in a D_2O atmosphere causes a jump in the scattering length density (from $1.5\times10^{-6}\text{Å}^{-2}$ to $2.5\times10^{-6}\text{Å}^{-2}$), due to the replacement of protons with deuterium in the ammonium groups (PAH->PAD). For the measurements at constant temperature, the solid line is a linear fit. The other line is a guide to the eye.

© 2004 WILEY-VCH Verlag GmbH & KGaA, Weinheim

Doing heating experiments at 100% r.h., one has to realize that the water vapor pressure increases with the temperature according to $p_V = b e^{-Q/RT}$ with $b = 76.8$ GPa and $Q = 42.03$ kJ/mol. Neutron reflectivity measurements of a (PSS/PAH)$_{10}$ film (prepared from 3 M NaCl solutions at room temperature) in a saturated D_2O-atmosphere at T=30, 40, 50, 60, 80, and 100°C are shown in Fig. 4(a).[13] The interference minima shift continuously to the right, indicating film thinning and water desorption. Careful data analysis reveals that the thickness decrease is accompanied by a slower than linear decrease of the scattering length density (cf. Fig. 4(b)), indicating not only water desorption, but also more efficient packing of some molecular groups.

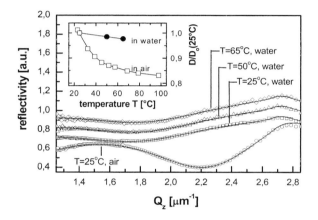

Figure 5. Optical reflectivity of a 32 layer pair PSS/PAH-film (prepared at room temperature from 3 M NaCl solution) on silicon. From bottom to top: in air, 25°C at ambient conditions, in H$_2$O at 25, 50 and 65°C. Symbols represent data points, solid lines corresponding fits. The reflection spectra are shifted for a better comparison. Inset: Normalized thickness (D/D$_0$(25°C)) vs. temperature for the 32 layer pair films in air (ambient conditions) and in water.

Even though the water vapor pressure increases due to heating, water molecules are expelled from the film, also its mass density increases. To explore the film shrinking at constant water chemical potential, the films were immersed in water, and their thickness is measured with optical reflectivity (cf. Fig. 5). On heating, again film shrinking is observed, yet it is substantially

© 2004 WILEY-VCH Verlag GmbH & KGaA, Weinheim

less than at in humid air, only 3%. With optical reflectivity, the thickness changes on temperature increase were measured. They take hours. The long time scale indicates that polymer rearrangement occurs, a fact which is consistent with the increased mass density of the heated films at 100% r.h.

The film shrinking and the increased mass density can be attributed to a force, which increases both the attraction between polymer segments, and repels water from the monomers. This fact, together with the pronounced temperature dependence[13] confirms that within a PSS/PAH multilayer, local interactions of hydrophobic nature exist and influence the polymer conformation.

Conclusion and Outlook

We investigated the self-organization of PSS at interfaces, and explored different effects: We identified and characterized the theoretically predicted polyelectrolyte brush phases.[10,11] We studied polyelectrolyte multilayer build-up as function of concentration and kind of salt, as well as temperature, and confirmed the increased layer thickness with increasing salt concentration.[12] We were interested in polyelectrolyte multilayers as humidity sensors, and studied swelling.[13] In all cases, we found evidence of a strong hydrophobic force, which showed the characteristic temperature behavior.

In diblock copolymer monolayers at the air/water interface, besides the polyelectrolyte brush phases we always find a flatly adsorbed PSS monolayer beneath the hydrophobic block.[10,11] Obviously, PSS experiences a strong attraction by a hydrophobic surface.

When the temperature[8] during polyelectrolyte multilayer build-up is varied between 5 to 40°C (at high salt conditions, which reduce range and amplitude of the electrostatic force), a thickness increase of up to 40% is observed. Simultaneously, the film/air roughness increases. Both the larger thickness and the roughening are indicative of an increased percentage of weak non-electrostatic bonds within the film. The temperature dependence of these effects is consistent with the hydrophobic force.

Studying the swelling of polyelectrolyte multilayers, we unintentionally probed the local interactions. At room temperature, the thickness and the water content increase fast (equilibrium attained in less than 1 s) and linearly with the partial water pressure (i.e. the r.h.). Therefore, the film can be seen as a mesh consisting of many voids, to which water molecules adsorb or from which they desorb. However, on heating at 100% r.h., the behavior is very different: the film

© 2004 WILEY-VCH Verlag GmbH & KGaA, Weinheim

shrinks while its mass density increases. It takes hours, until equilibrium is attained. Obviously, the polymer chains rearrange, increasing the packing efficiency. The hydrophobic segments are attracted to each other more strongly, and maximize their contact while the binding of water molecules is minimized.

We report very different experiments concerning the adsorption of PSS to hydrophobic and oppositely charged surfaces, as well as rearrangements within nanocomposites consisting partly of PSS. All observations are consistent with strong electrostatic forces. Yet for PSS, the hydrophobic force is obviously the most relevant of the secondary interactions, for other polyelectrolytes it may be different. There are other secondary interactions, like hydration forces, van der Waals interaction, specific forces, etc.[21] With the exception of some biological polymers and proteins, most polyelectrolytes have hydrophobic groups. To clarify the hierarchy of interactions during the self-assembly of polyelectrolytes much more experiments are necessary.

Acknowledgement

We thank Paul Simon and Karlheinz Graf for the help with the reflectivity experiments. The financial support of the DFG (He 1616/7-4) and the BMBF (03C0291C/5) are gratefully acknowledged.

[1] G. Decher, *Science* **277**, 1232-1237 (1997).

[2] M. Schönhoff, *Curr. Opin. Colloid Interface Sci.* **in print** (2003).

[3] R. R. Netz and J.-F. Joanny, *Macromolecules* **32**, 9013-9025 (1999).

[4] R. R. Netz, *J. of Phys. - Condensed Matter* **15**, s239-s244 (2003).

[5] S. S. Shiratori and M. F. Rubner, *Macromolecules* **33**, 4213-4219 (2000).

[6] H. Ahrens, H. Baltes, J. Schmitt, H. Möhwald, and C. A. Helm, *Macromolecules* **34**, 4504-4512 (2001).

[7] G. J. Fleer, M. A. C. Stuart, J. M. H. M. Scheutjens, T. Cosgrove, and B. Vincent, *Polymers at Interfaces* (Chapman and Hall, London, 1993).

[8] H. Lodish, A. Berk, S. L. Zipursky, P. Matsudaira, D. Baltimore, and J. Darnell, *Molecular Cell Biology*, 4th ed. (W.H. Freeman and Company, New York, 2000).

[9] S. Förster and M. Schmidt, *Advances in Polymer Science* **120**, 51-133 (1995).

[10] H. Ahrens, S. Förster, and C. A. Helm, *Macromolecules* **30**, 8447-8452 (1997).

[11] H. Ahrens, S. Förster, and C. A. Helm, *Phys. Rev. Lett.* **81**, 4172-4175 (1998).

[12] K. Büscher, H. Ahrens, K. Graf, and C. A. Helm, *Langmuir* **18**, 3885-3591 (2002).

© 2004 WILEY-VCH Verlag GmbH & KGaA, Weinheim

[13] J. Schmitt, D. Eck, P. Simon, and C. A. Helm, **subm.** (2003).

[14] H. Baltes, M. Schwendler, C. A. Helm, and H. Möhwald, *J. Colloid and Interface Science* **178,** 135-143 (1996).

[15] R. Sedev, R. Steitz, and G. H. Findenegg, *Physica B* **315,** 267-272 (2002).

[16] L. G. Parratt, *Phys. Rev.* **95,** 359 (1954).

[17] J. S. Pedersen and I. W. Hamley, *J. Appl. Cryst.* **27,** 36-49 (1994).

[18] O. V. Borisov, E. B. Zhulina, and T. M. Birshtein, *Macromolecules* **27,** 4795 (1994).

[19] A. Asnacios, R. Klitzing, and D. Langevin, *Colloids Surf. A* **167,** 189-197 (2000).

[20] H. Yim, M. Kent, A. Matheson, R. Ivkov, S. Satija, J. Majewski, and G. S. Smith, *Macromolecules* **33,** 6126-6133 (2000).

[21] J. N. Israelachvili, *Intermolecular and Surface Forces*, 2nd ed. (Academic Press, London, 1991).

[22] J. Schmitt, P. Mächtle, D. Eck, H. Möhwald, and C. A. Helm, *Langmuir* **15,** 3256-3266 (1999).

[23] M. Lösche, J. Schmitt, G. Decher, W. G. Bouwman, and K. Kjaer, *Macromolecules* **31,** 8893-8906 (1998).

[24] R. Kügler, J. Schmitt, and W. Knoll, *Macrom. Chem. Phys.* **203,** 413-419 (2002).

© 2004 WILEY-VCH Verlag GmbH & KGaA, Weinheim

Charge Density, Solvent Polarity and Counterion Nature Effects on the Solution Properties of Some Polycations

Stela Dragan, *[1] *Luminita Ghimici,*[1] *Christine Wandrey*[2]

[1]"Petru Poni" Institute of Macromolecular Chemistry, Aleea Grigore Ghica Voda 41 A, RO-6600 Iasi, Romania
E-mail: sdragan@ icmpp.tuiasi.ro
[2]Institute of Chemical and Biological Process Sciences, Swiss Federal Institute of Technology, CH-1015 Lausanne, Switzerland

Summary: The electrolytic conductivity and solution viscosity of some strong polycations, which possess the ammonium quaternary centers either attached to an acrylic macromolecular chain (polycation PDMAEMQ type), or in the backbone (polycation PCA type), have been measured in order to identify how charge density, polyelectrolyte concentration, solvent polarity as well as counterion nature influence the ultimate solution properties. The investigations reveal for all polycations an increase of the equivalent conductivity with decreasing concentration. The onset of the strong increase shifts to lower concentration if the charge density augments. Polyelectrolyte behavior was observed for PCA_5D_1 in the mixture of water/methanol for all solvent compositions employed and in the mixture of water/acetone up to 60 vol % of acetone content. At all experimental concentrations the reduced viscosity decreases with the dielectric constant of the mixed solvent. Furthermore, the specific interaction between several mono- and divalent counterions with some of these polycations was clearly proved by conductometric measurements.

Keywords: electrolytic conductivity; polycations; polyelectrolytes; polyion-counterion interactions

Introduction

Polyelectrolytes are increasingly important as advanced polymers in many fields including environmental technologies, biotechnology, and medicine.[1-2] However, their solution behavior and structure-properties interdependencies are far from being fully understood. Both intramolecular and intermolecular electrostatic interactions represent dominant factors influencing the properties of polyelectrolytes in solution.[3-5] The significance of intermolecular interactions has been suggested by Ise et al.[6,7] as well as Priel et al.[8] and was recently demonstrated and approved by viscometric experimental studies employing polyelectrolyte-surfactant complexes[9], stiff-chain polyelectrolytes[10], branched polyelectrolytes[11], as well as spherical architectures[12-14] as model systems. Viscosity studies are limited in their precision if the polymer concentration decreases, in particular, if the molar masses are

DOI: 10.1002/masy.200450708

not very high. On the other hand, electrolytic conductivity studies in aqueous solution, if carefully performed, can be extended up to concentrations far below the critical overlap concentration[5, 15] of low and medium molar mass polyelectrolytes only limited by the purity of the solvent and the onset of its self-dissociation. Therefore, electrolytic conductivity measurements are a powerful tool to study the polyion - counterion interactions dependent on the polyion's chemical structure, concentration, and the medium properties.[5, 16-18] It is well recognized that in solvents without added salt the electrostatic interaction between polyions and their counterions primarily depends on the polyion charge density,[19-21] though the influence of the chemical structure has been reported, too.[22]

In this study, the effect of charge density, polyelectrolyte concentration, as well as dielectric constant of the solvent on the viscometric behavior and the electrolytic conductivity of some polycations which possess the ammonium quaternary centers either attached to an acrylic macromolecular chain (polycation PDMAEMQ type) or in the backbone (polycation PCA type), have been analyzed in aqueous solution and aqueous mixtures. Reports on the electrolytic conductivity of polyelectrolytes in non-aqueous solution or in mixed solvents are relatively scarce, though can provide important information about electrostatic interactions.[23-26] In addition, specific interactions between some polycations of the PCA type and various counterions have been examined and will be discussed.

Experimental

Materials

Copolymers of the PDMAEMQ type with quaternary ammonium salt groups in the side chain, and with controlled quaternization degrees were synthesized by quaternization of poly(N,N-dimethylaminoethyl methacrylate) (PDMAEM) (M_w = 42340 g/mol and M_w/M_n = 1.98) with benzyl chloride (BC).[27] Their general structure is presented in Scheme 1.

Table 1 lists some characteristics of the copolymers. The charge density parameter in Table 1 is defined as $\xi = l_B/b =$, with b the average charge distance, $l_B = e^2/4\pi\varepsilon k T$ the Bjerrum length containing e the elemental charge, ε the dielectric constant of the solvent, k the Boltzmann constant, and T the temperature in K.[28,29]

© 2004 WILEY-VCH Verlag GmbH & KGaA, Weinheim

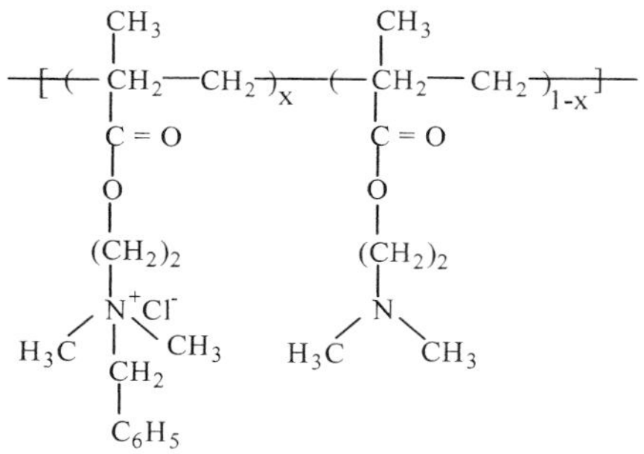

x = 0.5, polycation $PDMAEMQ_{50}$, x = 0.75, polycation $PDMAEMQ_{75}$,
x = 0.85, polycation $PDMAEMQ_{85}$

Scheme 1. Chemical structure of the PDMAEMQ type polyelectrolytes

Table 1. Characteristics of the quaternized PDMAEM samples.

Sample	Cl_i	Quaternization degree[a]	b[b]	ξ
	%	mol.-%	nm	
$PDMAEMQ_{50}$	8	50	0.50	1.42
$PDMAEMQ_{75}$	10.57	75	0.33	2.15
$PDMAEMQ_{85}$	11.45	85	0.29	2.43

[a] calculated on the basis of the ionic chlorine content;

[b] based on 0.25 nm for vinylic polymers.

The cationic polyelectrolytes of the PCA type were synthesized by polycondensation of epichlorohydrin (ECH) with dimethylamine (DMA) and N,N-dimethyl-1,3-diaminopropane (DMDAP) – polymer PCA_5, and primary amines with non-polar chains (hexyloxypropylamine - polycation PCA_5H_1 and decyloxypropylamine - polycation PCA_5D_1, according to the method presented in detail elsewhere.[30,31] Their general structures are presented in Scheme 2.

© 2004 WILEY-VCH Verlag GmbH & KGaA, Weinheim

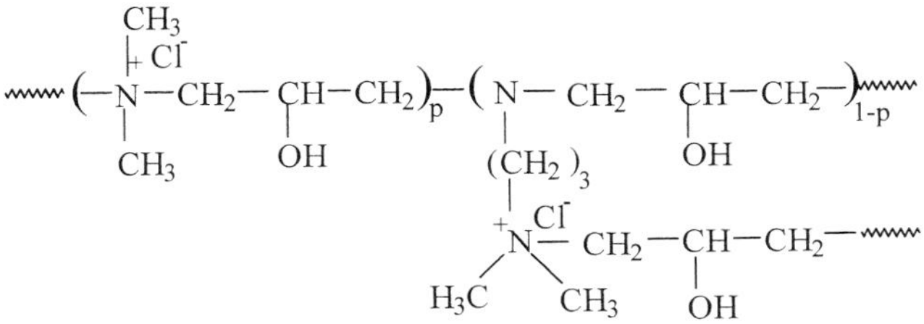

p = 0.95, polycation PCA$_5$

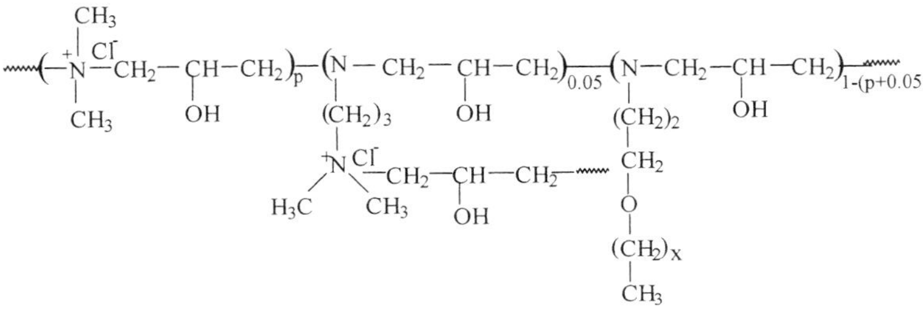

p = 0.94, x=5, polycation PCA$_5$H$_1$; p = 0.94, x=9, polycation PCA$_5$D$_1$

Scheme 2. Chemical structure of the PCA type polyelectrolytes

The definitions of the acronyms of these polycations are the followings: PC – polycation; A – asymmetrical amine; H, D – hydrophobic amine, the first index number designates mole percent of the polyfunctional amine whereas the last one stands for mole percent of the hydrophobic amine. All polycations were purified by ultrafiltration (Hollow-Fiber Concentrator CH2A, Amicon, USA) and solid polymers were obtained after vacuum freeze-drying (Beta 1-16, Christ, Germany).

Cationic polyelectrolytes of the PCA type were characterized by the content in ionic chlorine (determined by potentiometric titration with 0.02 N aqueous AgNO$_3$ solution) (Cl$_i$), total chlorine (determined by the combustion method – Schöniger technique) (Cl$_t$) and intrinsic viscosity, [η], in 1 M NaCl aqueous solution: Cl$_i$ = 21.67%, Cl$_t$ = 22.56%, [η]$_{1M\ NaCl}$ = 56.6 ml/g at 25 °C for PCA$_5$; Cl$_i$ = 19.82%, Cl$_t$ = 19.79%, [η]$_{1M\ NaCl}$ = 39.5 ml/g for PCA$_5$H$_1$; Cl$_i$ = 20.66%, Cl$_t$ = 21.03%, [η]$_{1M\ NaCl}$ = 45.0 ml/g for PCA$_5$D$_1$.

© 2004 WILEY-VCH Verlag GmbH & KGaA, Weinheim

Methods

Viscometric measurements of the polyelectrolyte dissolved in water, methanol, mixtures of water/methanol, or water/acetone were carried out at 25 ± 0.03 °C using an Ubbelohde type viscometer with automatic internal dilution (Viscologic TI1, SEMATech, France). All aqueous polymer solutions were prepared in double-distilled water before diluting with the organic solvent.

Conductivity measurements were performed with a model 712 conductometer (Metrohm, Herisau, Switzerland) using a conductivity cell with platinized electrode. All measurements were carried out at 20 °C $\pm$ 0.03 °C under nitrogen atmosphere. The polyelectrolyte was dissolved in highly purified deionized water from a Milli-Q PF (Millipore, Switzerland). The specific conductance of the water in the measuring vessel was always in the range 2 x 10^{-7} to 4 x 10^{-7} S cm^{-1}.

Results and discussion

Charge Density Effects

The reduced viscosity (η_{red}) – polymer concentration (c_p) profiles in water for the copolymers PDMAEMQ are illustrated in Figure 1. All copolymers exhibit polyelectrolyte behavior in water, that is, a typical upturn of the $\eta_{red} = f(c)$ plots at low concentration. The reduced viscosity values, at the same polymer concentration, are seen to increase in the order PDMAEMQ$_{50}$ < PDMAEMQ$_{75}$ < PDMAEMQ$_{85}$, that means with increasing charge density.

As the polymer concentration decreases the Debye length, $l_D = (4\pi N_A l_B \xi^{-1} c_p)^{-1/2}$ for polyelectrolyte solutions without added salt, increases resulting in an enhanced contribution of intermolecular repulsive interactions and, consequently, η_{red} increases. In terms of l_D, it also increases if the charge distance b decreases.

A more detailed discussion of the viscosity results can be performed on the basis of equation (1)[11], modified for $c_s \rightarrow 0$,

$$\eta_{red} = [\eta]_0 + k_H [\eta]_0^2 c_p + \frac{1}{160} m_l \left(\frac{4\pi e^2 N_A}{M_p \varepsilon \varepsilon_0 kT} \right)^{1/2} \frac{Z_{eff}^{5/2}}{c_p^{1/2}} \tag{1}$$

where the first two terms represent the Huggins equation. The third term considers polyelectrolyte behavior. Z_{eff} is the effective charge number per polymer, which was found to increase with decreasing ionic strength.[11]

© 2004 WILEY-VCH Verlag GmbH & KGaA, Weinheim

112

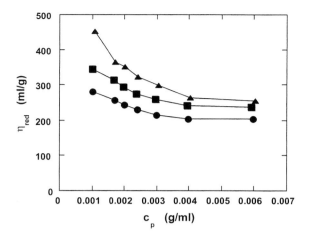

Figure 1. Reduced viscosity, η_{sp}/c, as a function of the concentration for polyelectrolytes with various charge distances in water: ▲ Q85, ■ Q75, ● Q50, T = 25 °C

The Odijk theory predicts for the critical overlap concentration c* independence of the charge density for highly charged polyelectrolytes, if the highly dilute concentration regime is reached.[32] The calculation according to $c^* = (N_A L^2 a)^{-1}$ yields 5.75×10^{-3} monomol/L (contour length L = 34 nm calculated from the molar mass of PDMAEM; monomer unit length a = 0.25 nm), or 1.52×10^{-3} g/ml, 1.45×10^{-3} g/ml, and 1.2×10^{-3} g/ml for PDMAEMQ$_{85}$, PDMAEMQ$_{75}$, and PDMAEMQ$_{50}$, respectively. Comparison of these values with the experimental data of Figure 1 identifies the experimental range to be above c* where no change of the counterion activity is expected.[22]

The conductivity investigations were performed by measuring the specific conductivity, κ, as a function of the polymer concentration, c_p, where c_p is the equivalent concentration of the polyelectrolyte in monomol/L. The specific conductivity considers the contribution of counterions and polyions to the current transport and depends on both the number of ions per unit volume and on their mobility. The equivalent conductivity, Λ, of the polyelectrolyte was calculated from equation 2:[16, 33-36]

$$\Lambda = \frac{\kappa - \kappa_0}{c_p} = f_c \left(\lambda_c^0 + \lambda_p \right) \tag{2}$$

where κ is the specific conductivity of the solution, and κ_0 is the specific conductivity of the solvent.

© 2004 WILEY-VCH Verlag GmbH & KGaA, Weinheim

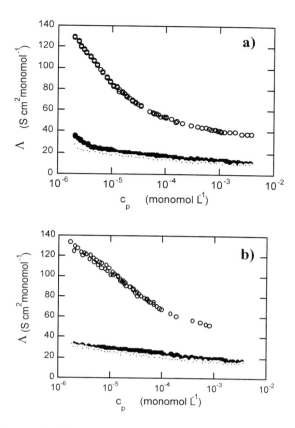

Figure 2. Equivalent conductivity, Λ, as a function of concentration for polycations of various charge density in various solvents: a) Q_{85}, b) Q_{50}; ($\circ$) water, ($\bullet$) water/methanol 1:1, ($\bullet$) methanol. T = 20 °C.

Figure 2a and Figure 2b show the concentration dependence of the equivalent conductivity in water for the two copolymers PDMAEMQ$_{85}$ and PDMAEMQ$_{50}$. The conductivity has been measured over three orders of magnitude of the polyion concentration, from 2×10^{-6} up to about 5×10^{-3} monomol/L.

The equivalent conductivity increases with decreasing concentration for all polycations and solvents. It seems that the onset of the strong increase in water is shifted to a lower concentration for a higher charge density. Furthermore, the equivalent conductivity was determined to be higher for the lower charge density, over the whole concentration range. This is expected according to Equation (2) with f_c higher for the lower charge density sample

© 2004 WILEY-VCH Verlag GmbH & KGaA, Weinheim

PDMAEMQ$_{50}$ if the equivalent conductivity is plotted vs. the equivalent concentration in monomol/L. No theory can explain the strong increase below a certain concentration quantitatively. However, an empirical model based on experimental findings has been proposed to explain this behavior.[15]

Solvent Polarity Effects

The environment that is the solution ionic strength and the solvent polarity also affects the electrostatic interaction between charged groups of the polyions and counterions. Extensive experimental data are available exploring the ionic strength influence. By comparison, reports on the electrolytic conductivity and viscometric behavior of polyelectrolytes in non-aqueous solution or in mixed solvents are relatively scarce,[23,26,37,38] though can provide important information about electrostatic interactions useful for practical activities. Therefore, the conductometric behavior of the PDMAEMQ polyelectrolytes was studied in methanol and solvent mixtures of water/methanol. The results presented in Figure 2a (PDMAEMQ$_{85}$) and Figure 2b (PDMAEMQ$_{50}$) clearly show differences by modifying the solvent quality; the equivalent conductivity decreases with the dielectric constant. To explain the solvent influence on the equivalent conductivity the change of the dielectric constant with its effects on the Bjerum length, l_B, and the Manning parameter, ξ, have to be considered. Both l_B and ξ increase if the dielectric constant of the solvent decreases yielding a lower f_c in Equation (2). As a consequence, less ions will participate in the current transport.

Also the viscosity of a polyelectrolyte solution depends on the solvent polarity, which affects the electrostatic interactions Varying the dielectric constant in the range of 78.5 to 33.3 the reduced viscosity, as a function of the polymer concentration, is shown for the polycation PCA$_5$D$_1$ dissolved in mixtures of water/methanol and water/acetone in Figure 3. The plots reveal that the reduced viscosity values increase with decreasing polymer concentration for all mixed solvents. This typical polyelectrolyte behavior, which is usually observed in salt-free aqueous solutions, occurs because the quaternary ammonium salt groups along the backbone can ionize in these polar solvent mixtures.

© 2004 WILEY-VCH Verlag GmbH & KGaA, Weinheim

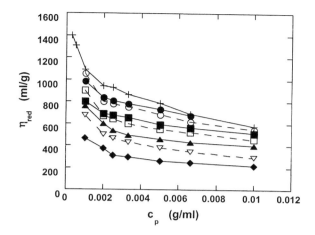

Figure 3. Influence of the dielectric constant on the reduced viscosity of PCA$_5$D$_1$ in water/methanol: (+) 100/0; (●) 75/25; (■) 50/50; (▲) 25/75; (◆) and water/acetone: (O) 75/25; (□) 50/50; (∇) 40/60.

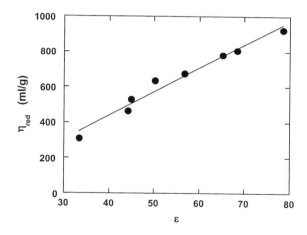

Figure 4. Reduced viscosity as a function of the dielectric constant, $c_p = 2.5 \times 10^{-3}$ g/mL.

For acetone, being a non-solvent for this polymer, the measurements also served to determine the acetone concentration above which the polyelectrolyte precipitates. Polyelectrolyte behavior of PCA$_5$D$_1$ was observed in water/acetone mixtures with acetone contents up to 60 vol %, whereas, polymer precipitation occurred at higher acetone concentration.

© 2004 WILEY-VCH Verlag GmbH & KGaA, Weinheim

It has been demonstrated that both Z_{eff} and the hydrodynamic radius r_H in Equation (1) increase with the dielectric constant[9]. This is in accordance with this study and is supported by plotting η_{red} vs. ε as it is illustrated in Figure 4 for $c_p = 2.5 \times 10^{-3}$ g/mL. Appropriate dependencies exist for the other concentrations. The overall contribution of the dielectric constant is more complicated since it affects according to Equation (1) the hydrodynamic radius, and Z_{eff} directly but has, additionally, a reciprocal influence.

Specific Interactions between Polyions and Counterions

This study was stimulated on the one hand by the fact that the most studies of counterion-polyion interactions were carried out in systems containing anionic polyelectrolytes and counterions; by comparison studies on cationic polyelectrolytes are relatively rare.[19,39-43] On the other hand the polyelectrolytes of the PCA type have proved efficiency in fields such as water purification,[44] formation of interpolyelectrolyte complexes,[45-47] or surface modification.[48,49] Since in these application areas low molar mass electrolytes can be present it was interesting to address this problem. Herein, selected conductometric results are presented. It is known that the electrical transport properties of polyelectrolyte solutions vary with the counterion type if low molar mass salts are added. Different strength of the interactions between these polymers and counterions was suggested. The theory predicts that the counterions associated with the polyion do not,[33] or only partially,[50] contribute to the conductivity whereas the free counterions behave as in a simple salt solution. Pursuing the variation of the molar conductivity of the polyelectrolyte (Λ_m) in different aqueous salt solutions as a function of the polyelectrolyte concentration for three polycations of the PCA type, the results plotted in Figure 5 and Figure 6 were obtained.

For polycations PCA_5 and PCA_5D_1, Figure 5 shows that the molar conductivity increases slowly with decreasing c_p in presence of univalent low molar mass salts. In the case of the SO_4^{2-} counterion, the curves exhibit a minimum between 1×10^{-3} unit mol/L and 3×10^{-3} unit mol/L. The minimum could be explained by the increase of the counterion-polyion interactions due to the capacity of the divalent anion to bind two cationic groups by intra- or interchain bridges. One may also observe that at the same polymer concentration, Λ_m decreases in the following sequence: $Cl^- > Br^- > I^- > SO_4^{2-}$, concluding the counterion binding increases in the same order. This sequence is similar to that found by other authors for different polycations.[39-43]

© 2004 WILEY-VCH Verlag GmbH & KGaA, Weinheim

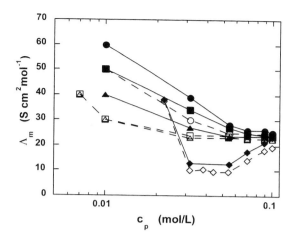

Figure 5. Influence of the counterion type on the concentration dependence of the molar conductivity (Λ_m) for PCA$_5$ (open symbols) and PCA$_5$D$_1$ (filled symbols) in: ($\circ$) and ($\bullet$) NaCl; ($\square$) and ($\blacksquare$) NaBr; ($\triangle$) and ($\blacktriangle$) NaI; ($\diamond$) and ($\blacklozenge$) Na$_2$SO$_4$.

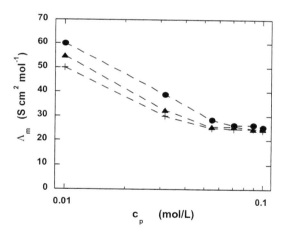

Figure 6. Influence of the polycation structure on the molar conductivity (Λ_m) in NaCl 10^{-3} M aqueous solution: ($\bullet$) PCA$_5$D$_1$; ($\blacktriangle$) PCA$_5$H$_1$; (+) PCA$_5$.

The increase of the polyion-counterion interactions for Cl$^-$ compared with I$^-$, i. e. with the decrease of the radius of the hydrated counterion, indicates that the counterion binding is not accompanied by appreciable dehydration and suggests, as assumed by other authors,[51] that the counterion binding in the halide series is non-specific. The counterion condensation

© 2004 WILEY-VCH Verlag GmbH & KGaA, Weinheim

phenomenon is specific for polyelectrolytes depending on their charge density[28, 29] and persists in the presence of low molar mass salts even at salt concentration up to 1 M.[52]

To gain information about the influence of the polycation structure on the counterion binding, the interaction between NaCl and three cationic polyelectrolytes differing by both their content of (N,N-dimethyl-2-hydroxypropyleneammonium) chloride units in the main chain and the length of the non-polar side chain was monitored, (Figure 6).

For comparable polyelectrolyte concentrations, the Λ_m values decrease in the order: $PCA_5D_1 > PCA_5H_1 > PCA_5$, i. e. with decreasing of the non-polar chain length and increasing the charge density. This might be taken as an indication that the presence of a non-polar side chain in the polyelectrolyte repeat unit causes the decrease of the polyion-counterion interactions.

Conclusions

Flexible highly charged cationic quaternary ammonium polyelectrolytes soluble not only in water but also in solvents and solvent mixtures with dielectric constant up to 30 have been synthesized. The study of both molecular and medium influences on the electrolytic conductivity and the reduced viscosity reveal interesting dependencies, which have not been reported before. Excluding the influence of molar mass the effect of the charge density has been approved on the reduced viscosity in water and the electrolytic conductivity in water, methanol, and mixtures thereof. A linear dependency of the reduced viscosity on the dielectric constant has been estimated for these flexible polyelectrolytes though this could not be quantified with the existing theoretical approaches yet.

The description of the electrolytic conductivity behavior remains very complex. Existing theoretical approaches need reconsideration concerning the significance of intermolecular interactions for the charge transport. Furthermore, the concentration, structure, and medium dependent changes of the polyion-counterion interaction need to be incorporated. Appropriate and more detailed investigations are in progress. The experimental results presented herein are seen to be useful to contribute to such approaches.

Acknowledgement

The EPFL is acknowledged for financial support.

© 2004 WILEY-VCH Verlag GmbH & KGaA, Weinheim

[1]   S. K. Tripathy, J. Kumar, H. S. Nalwa, Eds., *"Handbook of polyelectrolytes and their applications"*, Vol. 1-3, Am. Scientific Publishers, 2002.
[2]   H. Dautzenberg, W. Jaeger, J. Kötz, B. Philipp, C. Seidel, D. Stscherbina, *"Polyelectrolytes: Formation, Characterization, Application"*, Carl Hanser, München 1994.
[3]   M. Mandel in *Encyclopedia of Polymer Science and Engineering*, 2nd Ed., H. F. Mark, N. M. Bikales, C. G. Overberger, G. Menges, Eds., New York 1988, Vol. 11, p 739.
[4]   K. S. Schmitz, *"Macroions in Solution and Colloidal suspension"*, VCH, New York 1993.
[5]   C. Wandrey, D. Hunkeler, in *"Handbook of polyelectrolytes and their applications"*, S. K. Tripathy, J. Kumar, H. S. Nalwa, Eds., Am. Scientific Publishers, 2002, Vol. 2, p. 147-172.
[6]   J. Yamanaka, H. Matsuoka, H. Kitano, M. Hasegawa, N. Ise, *J. Am. Chem. Soc.* **1990**, *112,* 587.
[7]   J. Yamanaka, H. Araie, H. Matsuoka, H. Kitano, N. Ise, T, Yamaguchi, S. Saeki, M Tsubokawa, *Macromolecules* **1991**, *24*, 6156.
[8]   J. Cohen, Z. Priel, *J. Chem. Phys.* **1990**, *93*, 9062.
[9]   M. Antonietti, S. Förster, M Zisenis, J. Conrad, *Macromolecules* **1995**, *28*, 2270.
[10]   G. Brodowski, A. Horvath, A. Ballauff, M. Rehahn, *Macromolecules* **1996**, *29*, 6962.
[11]   M. Antonietti, A. Briel, S. Förster, *Macromolecules* **1997**, *30*, 2700.
[12]   M. Antonietti, A. Briel, S. Förster, *J. Chem. Phys.* **1996**, *105*, 7795.
[13]   F. Gröhn, M. Antonietti, *Macromolecules* **2000**, *33*, 5938.
[14]   M. Antonietti, A. Briel, F. Gröhn, *Macromolecules* **2000**, *33*, 5950.
[15]   C. Wandrey, *Langmuir* **1999**, *15*, 4069.
[16]   P. van Leeuven, R. F. M. Cleven, P. Valenta, *Pure Appl. Chem.* **1991**, *63*, 1251.
[17]   O. V. Davydova, A. N. Zelikin, S. I. Kargov, V. A. Izumrudov, *Macromol. Chem. Phys.* **2001**, *202*, 1361.
[18]   O. V. Davydova, A. N. Zelikin, S. I. Kargov, V. A. Izumrudov, *Macromol. Chem. Phys.* **2001**, *202*, 1368.
[19]   B. Boussouira, A. Ricard, R. Audebert, *J. Polym. Sci., Part B: Polym. Phys.* **1988**, *26*, 649.
[20]   W. F. Reed, S. Ghosh, G. Medjahdi, J. Francois, *Macromolecules* **1991**, *24*, 6189.
[21]   A. N. Zelikin, N. I. Akritskaya, V. A. Izumrudov, *Macromol. Chem. Phys.* **2001**, *202*, 3018.
[22]   C. Wandrey, D. Hunkeler, U. Wendler, W. Jaeger, *Macromolecules* **2000**, *33*, 7136.
[23]   M. Hara, Ed. *"Polyelectrolytes. Science and Technology"*, Marcel Dekker, New York 1993.
[24]   S. Dragan, L. Ghimici, *Polymer* **2001**, *42*, 2887.
[25]   L. Ghimici, S. Dragan, *Colloid Polym. Sci.* **2002**, *280*, 130.
[26]   Y. Nishiyama, M. Satoh, *Macromol. Rapid Commun.* **2000**, *21*, 174.
[27]   S. Dragan, I. Petrariu, M. Dima, *J. Polym. Sci., Polym. Chem. Ed.* **1981**, *19*, 2881.
[28]   G. S. Manning, *J. Chem Phys.* **1969**, *51*, 924.
[29]   G. S. Manning, *Ann. Rev. Phys. Chem.* **1972**, *23*, 117.
[30]   S. Dragan, L. Ghimici, *Angew. Makromol. Chem.* **1991**, *192*, 199.
[31]   S. Dragan, L. Ghimici, *Bull. Inst. Pol. Iasi* **1998**, *44*, 109.
[32]   T. Odijk, *Macromolecules* **1979**, *12*, 688.
[33]   G. S. Manning, *Biopolymers* **1970**, *9*, 1543.
[34]   J. R. Huzienga, P. F. Grieger, F. T. Wall, *J. Am. Chem. Soc.,* **1950**, *72*, 2636.
[35]   H. Eisenberg, *J. Polym. Sci.* **1958**, *30*, 47.
[36]   T. Kuruczev, B. J. Steel, *Pure Appl. Chem.,* **1967**, *17*, 149.
[37]   V. O. Aseyev, S. I. Klenin, H. Tenhu, *J. Polym. Sci., Part B: Polym. Phys.* **1998**, *36*, 1107.
[38]   G. R. Barraza, L. M. Pena, and H. E. Rios, *Polymer Internat.* **1997**, *42*, 112.
[39]   E. Prokopova, J. Stejskal, *J. Polym. Sci., Polym. Phys. Ed.* **1974**, *12*, 1537.
[40]   M. Satoh, M. Hayashi, J. Komiyama, T. Iijima, *Polymer* **1990**, *31*, 501.
[41]   H. E. Rios, L. N. Sepulveda, C. I. Gamboa, *J. Polym. Sci., Part B: Polym. Phys.* **1990**, *28*, 505.
[42]   H. E. Rios, C. Gamboa, G. Ternero, *J. Polym. Sci., Part B: Polym. Phys.* **1991**, *29*, 805.
[43]   J. Nagaya, A. Minakata, A. Tanioka, *Langmuir* **1999**, *15*,4129.
[44]   S. Dragan, A. Maftuleac, I. Dranca, L. Ghimici, T. Lupascu, *J. Appl. Polym. Sci.* **2002**, *84*, 871.
[45]   S. Dragan, D. Dragan, M. Cristea, A. Airinei, L. Ghimici, *J. Polym. Sci., Part A: Polym. Chem.* **1999**, *37*, 409.
[46]   S. Dragan, M. Cristea, *Eur. Polym. J.* **2001**, *37*, 1571.
[47]   S. Dragan, M. Cristea, *Polymer* **2002**, *43*, 55.
[48]   S. Dragan, S. Schwarz, K.-J. Eichhorn, K. Lunkwitz, *Colloids and Surfaces A: Physicochem. Eng. Aspects* **2001**, *195*, 243.
[49]   S. Dragan, S. Schwarz, *Progr. Colloid Polym. Sci.* **2003**, *122*, 8.
[50]   N. J. Yoshida, *Chem. Phys., Lett.,* **1982**, *90*, 207.
[51]   M. Satoh, T. Kawashima, J. Komiyama, T. Iijima, *Polym. J.* **1987**, *19*, 1191.
[52]   G. S. Manning, *Ber. Bunsenges. Phys. Chem.* **1996**, *100*, 909.

© 2004 WILEY-VCH Verlag GmbH & KGaA, Weinheim

Macromol. Symp. **2004**, *211*, 121-133

The Stability of Polyelectrolyte Complex Systems of Poly(diallydimethyl-ammonium chloride) with Different Polyanions

Mandy Mende,[1] *Heide-Marie Buchhammer,*[1] *Simona Schwarz,*[1] *Gudrun Petzold,*[1] *Werner Jaeger*[2]

[1]Institute of Polymer Research Dresden, Hohe Str. 6, 01069 Dresden, Germany
E-mail: mende@ipfdd.de
[2]Fraunhofer Institute for Applied Polymer Research, Geiselbergstr. 69, 14476 Golm, Germany

Summary: Polyelectrolyte complex (PEC) dispersions formed by fast ionic exchange reaction between polyelectrolytes (PELs) bearing oppositely charged groups have been characterized by turbidity and dynamic light scattering. The stability of the formed PEC dispersions, the average hydrodynamic particle size d_h and the also determined polydispersity index PI of the PEC particles were of special interest. In this study poly(diallyldimethylammonium chloride) (PD) of different molar mass is used in combination with copolymers of maleic acid and α-methylstyrene (P(MS-αMeSty)) as well as propene (P(MS-P)), with a copolymer of acrylamide and sodium acrylate (PR2540^{UD}) and poly(styrene-p-sodium sulfonate) (NaPSS).
A strong dependence of the PEC stability on the molar charge ratio n_-/n_+, the mixing conditions and especially on the chemical structures of the used polyanions was found. If the employed polyanion (PA) have a π-system (phenyl-) in the polymer chain a higher tendency to instability results. The average hydrodynamic particle sizes d_h and the polydispersity indices PIs of the formed stable PEC dispersions determined by dynamic light scattering were strongly influenced by the mixing conditions of the PECs. If we are able to prepare stable complexes where the isoelectric point is exceeded during complex formation which is possible for PD/P(MS-P) and PD/PR2540^{UD} systems a distinct increase of d_h and decrease of PI is observed. The polymer concentration C_P and the chain length L of the used PELs have a minor effect on d_h and PI.

Keywords: dispersions; dispersion stability; dynamic light scattering; polyelectrolyte complex; polyelectrolytes

Introduction

It is well known that oppositely charged polyions (polycation, PC and polyanion, PA) dissolved in aqueous solution form polyelectrolyte complexes (PEC).[1-3] The following three different types can be distinguished[1]: (i) The so-called soluble complexes, which contain small PEC aggregates and therefore they form macroscopically homogeneous systems. (ii) Turbid

© 2004 WILEY-VCH Verlag GmbH & KGaA, Weinheim

DOI: 10.1002/masy.200450709

colloidal systems with suspended complex particles in the transition range to phase separation. (iii) Phase separation by precipitation of PEC-polysalt.

It depends on several parameters, such as the chain length L of the employed polyions, their chemical structures as well as the type and distribution of the charged groups along the polymer chains of what type ((i), (ii) or (iii)) a PEC is. Beside these structural influences polymer concentration C_P, molar ratio of mixed anionic to cationic charges n_-/n_+, ionic strength I, pH of the medium etc. play an important role.

It is a well-established fact that stable PECs with no phase separation during complexation can be formed under certain conditions, i.e. polyelectrolytes with weak ionic groups and significant different molecular weights at non-stoichiometric mixing ratios.[4-8] These PECs consist of a long host molecule and shorter sequentially attached guest polyions of opposite charge as comprehensively studied by the groups of Tsuchida and Kabanov.[9,10] In dependence on the ionic sites at the polyelectrolyte chains and the mixing conditions the formed stable PECs contain single stranded hydrophilic and double stranded hydrophobic sections. PECs of high hydrophobicity result in supermolecular structures and precipitate.

As shown and discussed in different studies the formed PEC particles consist of a neutralized, hydrophobic core and a shell of the excess component which stabilizes the particles against further coagulation.[11-13]

In connection with the use of the PEC particles as flocculating agent[14, 15] or as carrier for dyes or solved organic molecules from the waste water,[16] we investigated the stability and the behavior of PECs in more detail.[17]

In this study we want to compare the properties of four PEC systems, such as turbidity, particle size and their distribution, in dependence on the molar charge ratio n_-/n_+, the polymer concentration C_P and the initial polycation concentration C_0^{PC}, respectively, the chain length L of the PELs and the types of the functional groups of the polyanions. Additional the chemical structures of the employed polyions and the mixing conditions (starting solution) were of special interest.

Experimental part

Materials

The used cationic poly(diallyldimethylammonium chloride) (PD) has two different molecular weights, $M_w = 5 \cdot 10^3 \, g/mol$ and $M_w = 2.9 \cdot 10^5 \, g/mol$. The first one was purchased from Aldrich and used without further purification. The second one was synthesized by free radical

© 2004 WILEY-VCH Verlag GmbH & KGaA, Weinheim

polymerization in aqueous solution, purified by ultrafiltration and characterized as described in.[18]

Poly(styrene-p-sodium sulfonate)s (NaPSS) with molecular weights of $7 \cdot 10^4$ g/mol and 10^6 g/mol were also purchased from Aldrich and used as obtained.

Poly(acrylamid-co-sodium acrylate) (PR2540^{UD}), a commercial product (PRAESTOL®) was obtained by Degussa.Stockhausen GmbH & Co. KG (Krefeld, Germany). The molecular weight was about $M_W = 14 \cdot 10^6$ g/mol and the charge density was 40 wt.-% (according to the manufacturer). By means of ultrasonic degradation the molecular weight of the polymer has been decreased to $5 \cdot 10^5$ g/mol.[17, 19] The effective charge density determined by PEL-titration was 28 mol-% (33 wt.-%) before as well as after ultrasonic degradation.

Poly(maleic acid-co-propene) (P(MS-P)) and poly(maleic acid-co-α-methylstyrene) (P(MS-αMeSty)) were prepared by hydrolysis of the corresponding anhydrides. The pH-values of the solutions were adjusted to 6. These anhydrides were obtained from Leuna GmbH (Germany) with $M_W = 5 \cdot 10^4$ g/mol for the propene-copolymer and $M_W = 2.5 \cdot 10^4$ g/mol for the α-methylstyrene-copolymer. More characteristic parameters of the used PELs are summarized in Table 1 where P_W is the degree of polymerization, L is the contour length of the polymer chain and b is the average distance of charges along the polymer chain.

Table 1. Characteristic parameters of used PELs

PEL	M_W g/mol	P_W	L nm	b nm
PD	5 000 290 000	31 1800	15.5 900	0.5
PR2540^{UD}	500 000	6400	1970	1.1[a]
NaPSS	70 000 10^6	340 4850	85 1212	0.25
P(MS-P)	50 000[c]	357	178	0.5[d]
P(MS-αMeSty)	25 000[c]	116	58	0.5[d]

[a] calculated concerning 28 mol-% acrylate
[b] before ultrasonic degradation (UD) M_W was ca. 14x10^6 g/mol
[c] is correlated to the anhydride
[d] approximately the half of the functional groups of the copolymers are available as salt at pH ≈ 6

Preparation of Polyelectrolyte Complexes

The PECs were prepared by combination of equal amounts (50 ml) of aqueous solutions of oppositely charged PELs. These solutions were obtained by dilution of definite amounts of

© 2004 WILEY-VCH Verlag GmbH & KGaA, Weinheim

stock PEL solutions, $C_{PD} = 1.62\,g/l$, $C_{NaPSS} = 2.06\,g/l$, $C_{PR\,2540} = 1\,g/l$, $C_{P(MS-\alpha MeSty)} = 2.2\,g/l$ and $C_{P(MS-P)} = 1.4\,g/l$ in Millipore water. Concerning the charge contents all PEL solutions were characterized by PEL titration carried out by means of a PCD 02 particle charge detector (Mütek GmbH, Germany).

The initial concentration of polycation solution (C_0^{PC}) was kept constant for all prepared PECs within a complex series, while the amount of anionic charges corresponds to the desired n_-/n_+-ratio. n_-/n_+ is the calculated ratio of charges. Under continuous stirring one PEL solution was added to the used starting PEL solution with a flow rate of 0.2 l/h. In the experiments PC solutions as well as PA solutions were used as starting solution. Detailed information is given in Table 2. It should be mentioned that C_0^{PC} is the concentration of the polycation solution at the beginning of complex fomation ($v_0^{PC} = 50\,ml$). At the end of the complex formation the volume of the formed complex dispersion is twice the volume of the separate initial solutions. The polymer concentration C_P is relatd to the end state of the complex formation. Hence the value of C_P can be lower than C_0^{PC} in spite of that the amount of polycation as well as the amount of polyanion is included in C_P.

Table 2. Concentrations and molar ratios of formed PECs

PEC system		M_W^{PC}	M_W^{PA}	C_0^{PC}	n_-/n_+	C_P
		g/mol	g/mol	mmol/l		mmol/l
PD/NaPSS-I	a	5 000	10^6	4	0.4-2.0	2.8-6.0
PD/NaPSS-II	a b	290 000	70 000	4	0.4-2.0	2.8-6.0
PD/NaPSS-III	a b	5 000	70 000	4	0.4-2.0	2.8-6.0
PD/NaPSS-IV	a b	5 000	70 000	1	0.4-2.0	0.7-1.5
PD/PR2540^{UD}	a b	290 000	500 000	1.5	0.4-2.0	1.0-2.2
PD/P(MS-P)	a b	290 000	50 000	1.5	0.5-1.2	1.125-1.65
PD/P(MS-αMeSty)	a b	290 000	25 000	1.5	0.5-1.2	1.125-1.65

a...starting solution: PC (polycation)
b...starting solution: PA (polyanion)

© 2004 WILEY-VCH Verlag GmbH & KGaA, Weinheim

After complete addition the mixtures were still stirred for 10 minutes. Two hours later and after a storage time of 2 days the PEC dispersions were characterized.

Characterization of the Complex Dispersions

Measurements of optical density (OD) were used to characterize the stability of the PECs. The values of OD were obtained with a Lambda 900 UV/VIS/NIR spectrometer (Perkin-Elmer, UK). All measurements were made at $\lambda = 500$ nm. Deionized water was used as reference.

Dynamic light scattering measurements were performed by Zetasizer 3000 (Malvern Instruments, UK). The instrument was equipped with a monochromatic coherent 10 mW Helium Neon laser ($\lambda = 633$ nm) as light source. The light scattered by particles is recorded at an angle of 90°. The analysis of the autocorrelation function $g^{(2)}(\tau)$ was done automatically to yield the mean diffusion coefficient D_T. Thereafter z-average hydrodynamic diameter d_h of PEC particles was calculated by the Stokes-Einstein equation.

To get information about size distribution of PEC particles the polydispersity index (PI) was also included in the interpretation. PI is defined as

$$PI = \frac{2\,c}{b^2} \tag{1}$$

where b and c are the coefficients of an infinite series in powers of the delay time as which the logarithm of the autocorrelation function can be expressed. This is a dimensionless measure of the broadness of the distribution.

In addition all PECs were characterized by PEL titration (PCD 02, Mütek GmbH, Germany) in order to determine the isoelectric points $(n_-/n_+)_{IP}$ within a complex series.

Discussion

All investigated PECs were characterized by their charge excess. The determined isoelectric points of the PEC series are in the range of $0.85 \leq (n_-/n_+)_{IP} \leq 0.95$. The difference to $n_-/n_+ = (n_-/n_+)_{IP} = 1$ can be explained by the adsorption of the PC at the reaction vessel wall. We used glass vessels in order to observe the PEC formation visual. At the isoelectric point complete precipitation of all these formed PECs is observed. In Scheme 1 we summarize the general observations and in the next part we will discuss them in conjunction with the turbidity (OD_{500}).

At first let us look for the system PD/P(MS-αMeSty). In spite of low polymer concentrations ($C_0^{PC} = 1.5\,\text{mmol}/1$) non of the complexes within the series is absolute stable. Also at molar

© 2004 WILEY-VCH Verlag GmbH & KGaA, Weinheim

mixing ratios of charges, which guarantees a high charge excess, we found partial flocculation of the PECs. That means that precipitation occurs besides the formation of turbid supernatants and stable dispersions, respectively, which were stable for few weeks.

For the next PEC system PD/NaPSS stable complexes were found as long as the starting solution is in excess, independent if PC or PA is the starting solution, independent on the initial polymer concentration of the PC C_0^{PC} and independent on the chain length L of the employed PELs. If the starting solution is deficient at the end of complex formation, that means that we exceed the isoelectric point $(n_-/n_+)_{IP}$ during complex formation, partial precipitation is observed and the supernatants form dispersions as in the case of PD/P(MS-αMeSty) mentioned above.

Scheme 1. Stability of formed PECs

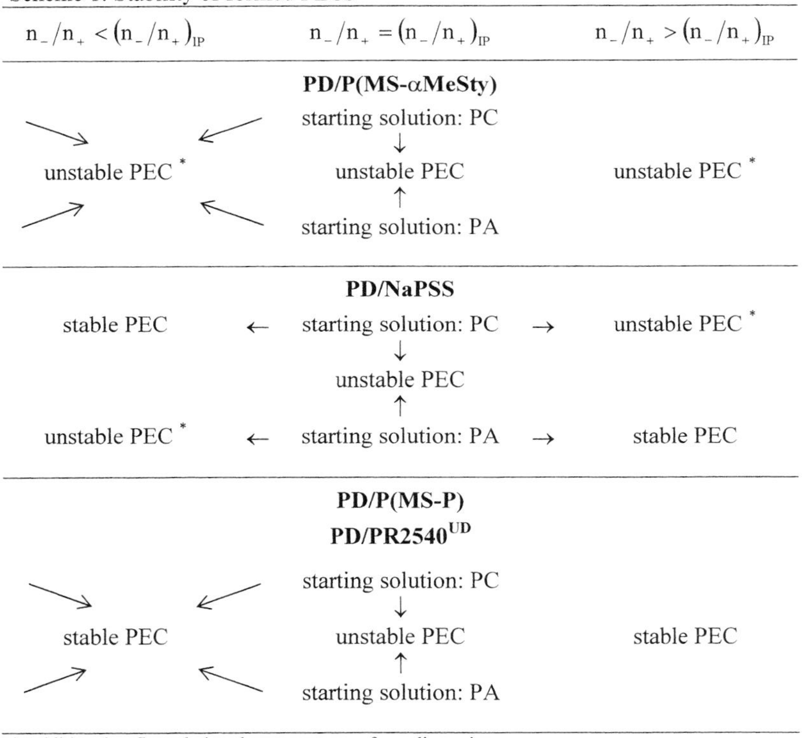

$n_-/n_+ < \left(n_-/n_+\right)_{IP}$	$n_-/n_+ = \left(n_-/n_+\right)_{IP}$	$n_-/n_+ > \left(n_-/n_+\right)_{IP}$
	PD/P(MS-αMeSty)	
	starting solution: PC	
	↓	
unstable PEC *	unstable PEC	unstable PEC *
	↑	
	starting solution: PA	
	PD/NaPSS	
stable PEC ←	starting solution: PC →	unstable PEC *
	↓	
	unstable PEC	
	↑	
unstable PEC * ←	starting solution: PA →	stable PEC
	PD/P(MS-P)	
	PD/PR2540^{UD}	
	starting solution: PC	
	↓	
stable PEC	unstable PEC	stable PEC
	↑	
	starting solution: PA	

* Additional to flocculation the supernatants form dispersions.

At the higher polymer concentrations the stability/instability is reflected by the values of turbidity (OD_{500}) in some degree as expected. But for the interpretation of the OD_{500} it should

© 2004 WILEY-VCH Verlag GmbH & KGaA, Weinheim

be taken into consideration, that the optical density is influenced by different parameters of the light scattered complex particles such as shape, number, size and polydispersity.

Figure 1a and b show the turbidity of the PEC system PD/NaPSS at $C_0^{PC} = 4\,mmol/l$ in dependence on the molar mixing ratio of the charges n_-/n_+ and the chain length L of the PELs. Moreover the starting solution was varied. In figure 1a PA was added to PC and the order of addition was changed in figure 1b.

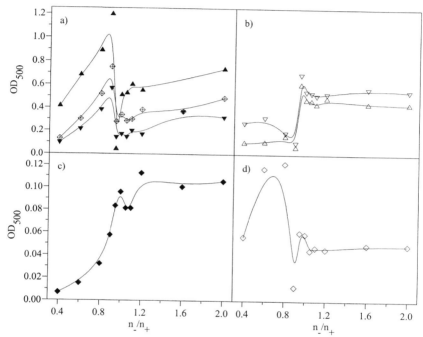

Figure 1. Optical density (OD$_{500}$) of PD/NaPSS complexes as a function of the molar charge ratio n_-/n_+; a) PC is starting solution, $C_0^{PC} = 4\,mmol/l$, ⊕ = PD/NaPSS-Ia, ▲ = PD/NaPSS-IIa, ▼ = PD/NaPSS-IIIa; b) PA is starting solution, $C_0^{PC} = 4\,mmol/l$, △ = PD/NaPSS-IIb, ▽ = PD/NaPSS-IIIb; c) PC is starting solution, $C_0^{PC} = 1\,mmol/l$, ◆ = PD/NaPSS-IVa; d) PA is starting solution, $C_0^{PC} = 1\,mmol/l$, ◇ = PD/NaPSS-IVb.

As presented in figure 1a the optical density increased in the range of $n_-/n_+ < (n_-/n_+)_{IP}$ with increasing n_-/n_+ for all the stable complexes. Near the isoelectric point a maximum is reached. Beyond the isoelectric point we found partial flocculation of the formed PECs. The OD$_{500}$-values of these dispersions are lower than for complete stable PECs normally. In our case the

© 2004 WILEY-VCH Verlag GmbH & KGaA, Weinheim

values are of the same magnitude as the OD_{500}-values for the stable complexes with $n_-/n_+ \ll (n_-/n_+)_{IP}$ of the same complex series because the polymer concentration C_P increases within the complex series. With increasing of n_-/n_+ in the range of $n_-/n_+ > (n_-/n_+)_{IP}$ the turbidity only slightly increases in contrast to the range of $n_-/n_+ < (n_-/n_+)_{IP}$.

If we compare complexes with the same n_-/n_+-ratio we could see that the turbidity rises with increasing chain length of the employed PELs. The PEL used as starting solution (PC) has a stronger effect on the magnitude of the turbidity as the added component (PA).

The behavior of the turbidity of the PD/NaPSS PECs at $C_0^{PC} = 4\,\mathrm{mmol}/\mathrm{l}$ with the polyanion as starting solution is shown in Figure 1b. Here we have a clear difference of the OD_{500}-values between the complete stable PEC dispersions and supernatants of unstable PEC. We did not found a pronounced maximum near $(n_-/n_+)_{IP}$.

The complexes of the series PD/NaPSS-IV presented in Figure 1c and d were prepared at $C_0^{PC} = 1\,\mathrm{mmol}/\mathrm{l}$. Hence the turbidity of the PECs is essentially lower than the turbidity at higher polymer concentrations. Surprisingly the OD_{500}-values of the supernatants of unstable PECs with $n_-/n_+ \neq (n_-/n_+)_{IP}$ are distinctly higher than the values of the stable PECs in most cases. This cannot be explained in each case by the increase of C_P within the complex series. But the OD_{500}-values increase over time slowly for the same n_-/n_+ in contrast to the stable PECs, which is admittedly not shown in the Figure. This behavior indicates aggregation and consequently instability.

The next two PEC systems, PD/P(MS-P) and PD/PR2540^{UD}, form at $C_0^{PC} = 1.5\,\mathrm{mmol}/\mathrm{l}$ stable PECs in the ranges where the starting solution is in excess as well as in deficient. Only at $n_-/n_+ = (n_-/n_+)_{IP}$ complete unstable PECs are observed because of the absence of charge excess.

As long as the starting solution is in excess OD_{500}-values are approximately constant and in a range of magnitude as stable PD/NaPSS PECs at $C_0^{PC} = 1\,\mathrm{mmol}/\mathrm{l}$ as seen in Figure 2. Only near the isoelectric point a raising is observed. If the isoelectric point is exceeded during complex formation a pronounced increase of the OD_{500}-values take place within a complex series. This increase correlates with particle size measurements which we will discuss later.

It should be noticed that the OD_{500}-values decrease with an increase of the charge excess if isoelectric point is exceeded during complex formation, that means with decrease of n_-/n_+ for the used PC in excess and with increase of n_-/n_+ for the employed PAs in excess. Whereas the decrease of the turbidity is pronounced for the increased excess of the PC as for the used PAs which could be an effect of the polymer concentration C_P. The polymer concentration increases with n_-/n_+ within a complex series.

© 2004 WILEY-VCH Verlag GmbH & KGaA, Weinheim

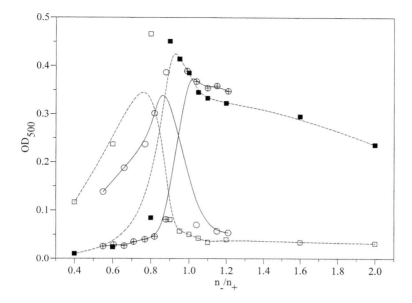

Figure 2. Optical density (OD_{500}) of the PEC systems which were stable at all n_-/n_+-ratios (except PECs at $(n_-/n_+)_{IP}$) as a function of the molar charge ratio n_-/n_+; ■ = PD/PR2540^{UD}-a; □ = PD/PR2540^{UD}-b; ⊕ = PD/P(MS-P)-a; ○ = PD/P(MS-P)-b; C_0^{PC} =1.5 mmol/l for PD/PR2540^{UD} and C_0^{PC} =1 mmol/l for PD/P(MS-P).

If we compare the OD_{500}-values of complexes with the same n_-/n_+ und hence the same C_P in dependence of the starting solution so we found higher turbidity if the $(n_-/n_+)_{IP}$ is exceeded during complex formation.

In the next discussion part we will study the average hydrodynamic particle sizes d_h and their distributions of all formed stable PECs. Figure 3a and Figure 3b show the measured particle sizes of all stable PECs of PD/NaPSS, PD/P(MS-P) and PD/PR2540^{UD} systems in dependence on n_-/n_+, the polymer concentration C_P and the initial concentration of the PC C_0^{PC}, respectively, and in dependence on the PEL chain length L. In order to interpret the value d_h better we also take into account in our analyses the so-called polydispersity index PI, presented in Figure 3c and Figure 3d, as a measure of the size distribution. From mathematical point of view the PI describes the deviation between the determined autocorrelation function in photon correlation spectroscopy and the adjusted correlation

© 2004 WILEY-VCH Verlag GmbH & KGaA, Weinheim

function. In practice dispersions with PI-values between 0.03 and 0.06 are called as monodisperse. A narrow distribution is available at values between 0.1 and 0.2. If the polydispersity index is in the range of $0.25 \leq PI \leq 0.5$ the distribution is broad and at values above 0.5 the measured result is not analyzable and indicates a broad distribution of indefinite particle shapes.[20]

Let us look at the d_h-values of the stable PD/NaPSS PECs. The average hydrodynamic particle sizes d_h are approximately constant within a complex series independent on the molar charge ratio n_-/n_+. Above all it is found for PA as starting solution. If PC is the starting solution a very low increase is observed like for OD_{500} too. The influence of the chain length L of the used PELs on d_h is not fairly clear. The most significant effect of the measured average hydrodynamic size d_h has the initial concentration of PC C_0^{PC}. The d_h-values for PECs at $C_0^{PC} = 1\,mmol/l$ are distinctly lower than that at $C_0^{PC} = 4\,mmol/l$. The behavior of the accompanying PIs is analogous. Essentially by decreasing the initial concentration C_0^{PC} the PIs decreased also.

In the range of $n_-/n_+ < (n_-/n_+)_{IP}$, if PC is the starting solution, the PIs decreased within a complex series with increasing n_-/n_+ normally. If PA is the starting solution $(n_-/n_+ > (n_-/n_+)_{IP})$ the change of the PIs within a complex series is less pronounced. Only near $(n_-/n_+)_{IP}$ the PIs decreased more distinctly with decrease of n_-/n_+. However the PIs of all PD/NaPSS PECs are high.

A comparison of the d_h of the PEC systems which form stable PECs at all molar charge ratios n_-/n_+ within a complex series, except of PEC with $n_-/n_+ = (n_-/n_+)_{IP}$, is shown in Figure 3b and the belonging at PIs in Figure 3d.

For both systems, PD/PR2540^{UD} and PD/P(MS-P), we found distinct higher d_h-values (ca. 250 – 300 nm) if isoelectric point is exceeded during complex formation. It is the case in the range of $n_-/n_+ < (n_-/n_+)_{IP}$ if PA is starting solution and in the range of $n_-/n_+ > (n_-/n_+)_{IP}$ if PC is starting solution. Simultaneously we observe very low PIs of such complexes. They are all in the range of monodisperse dispersions.

In contrast to this the average particle size d_h of PECs formed without exceeding the isoelectric point are about 100 – 150 nm lower at the same n_-/n_+ and in the range of stable PD/NaPSS PECs for the same C_0^{PC} $(= 1\,mmol/l)$. In the range of starting solution excess the PIs of stable PD/NaPSS PECs decrease in direction to isoelectric point, that means for PC as starting solution with increase of n_-/n_+ and for PA as starting solution with decrease of n_-/n_+ up to $(n_-/n_+)_{IP}$.

© 2004 WILEY-VCH Verlag GmbH & KGaA, Weinheim

Now we will give an explanation for the behavior of the average particle size d_h connected with polydispersity index PI. As long as the PEC formation stops if the starting solution is in excess the determined average particle size of the PEC particles and the PI belongs to it are definite by electrostatic interactions and kinetic. Because of the large size of the polyions in contrast to the small counterions, the polyions will react preferentially with the oppositely charged polyions in the neighborhood. So that bigger particles are formed initially beside smaller particles and hence a higher polydispersity is observed. The stabilization of the PEC

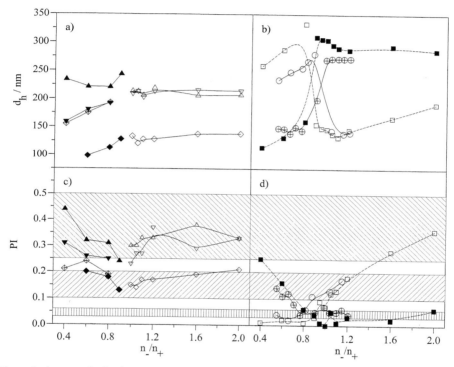

Figure 3. Average hydrodynamic particle size d_h and polydispersity index (PI) of the studied stable PEC dispersions in dependence of the molar charge ratio n_-/n_+; a) and c) PD/NaPSS systems, b) and d) PD/PR2540[UD] and PD/P(MS-P) systems; the symbols are the same as in Figure 1 and Figure 3; the marked fields for PI means \\\\\ broad distribution, ///// narrow distribution and ||||| monodisperse distribution.[19]

particles is caused by the excess PEL. If n_-/n_+-ratio approaches $(n_-/n_+)_{IP}$ electrostatic interaction will decrease and hydrophobic interaction will increase. Hence the probability of aggregation of PEC particles formed before will rise and the number of smaller PEC aggregates will decrease. Lower PIs result even though a drastic change in average hydrodynamic particle size d_h is not observed. In this moment where $(n_-/n_+)_{IP}$ is exceeded during PEC formation and hence starting solution is deficient in the mixture the PEC aggregates formed earlier were stabilized by adsorption of the excess PEL again. A further aggregation is prevented. But caused by the formed shell of the excess PEL onto the surface of the PEC aggregates formed before we determine a distinct higher average hydrodynamic particle size d_h and no change of PI.

The ability of PEC systems to form stable PECs is closely connected with hydrophobic interactions. If hydrophobic interactions play a dominant role a tendency to instability results. We studied two PEC systems (PD/P(MS-αMeSty) and PD/NaPSS) in which one of the used PELs contains a π-system. This π-system causes stronger hydrophobic interactions which means a stronger addiction to aggregation. But as in the case of NaPSS a functional group is bound at the π-system hydrophobic interactions play a dominant role only beyond the isoelectric point (and at $(n_-/n_+)_{IP}$) caused by the structure of the formed PECs. In contrast to this partially flocculation is found for all n_-/n_+-ratios in the case of P(MS-αMeSty) ($n_-/n_+ = (n_-/n_+)_{IP}$: complete flocculation) because of the absence of functional groups at the π-system. We observed no distinct effect on complex formation by weak or strong functional groups of the employed PELs.

Conclusion

Dynamic light scattering and turbidity have been used to study the effects of polyanion chain length L, initial polymer concentration of the polycation C_0^{PC}, molar charge ratio n_-/n_+, mixing conditions and chemical structure of the used polyanions on the stability of PEC dispersions and on properties of formed PEC particles such as average hydrodynamic particle size d_h and polydispersity index PI as a measure of particle size distribution.

It has been found that PEC systems which include a PA with a π-system (phenyl-) at the polymer chain (PD/P(MS-αMeSty) and PD/NaPSS) show a higher tendency towards instability. PD/P(MS-αMeSty) are not able to form complete stable PEC dispersions. The system PD/NaPSS forms stable PEC dispersions as long as the starting solution is in excess, that means stable PECs for PC as starting solution as long as $n_-/n_+ < (n_-/n_+)_{IP}$. If PA is employed as starting solution PECs with $n_-/n_+ > (n_-/n_+)_{IP}$ are stable. The determined average hydrodynamic particle sizes of the stable PD/NaPSS PECs are all in the range of about 150 – 250 nm if $C_0^{PC} = 4\,mmol/l$. It seems that the average d_h-value is determined by the chain

© 2004 WILEY-VCH Verlag GmbH & KGaA, Weinheim

length L of PEL which is in excess in the mixture. All of them have a very high polydispersity indices (PIs). If C_0^{PC} is reduced to 1 mmol/l the d_h-values are lower than 150 nm. Simultaneously the determined PI decreased also to the range of narrow distribution.

In contrast to the PEC systems mentioned above two of the studied systems, (PD/P(MS-P) and PD/PR2540^{UD}), form stable dispersions even if the starting solution is deficient in the mixture. That means that during complex formation the isoelectric point is exceeded. It is remarkable that the d_h increases by about 80 – 125 nm in contrast to these PECs which were formed at the same n_-/n_+ but under the condition that the starting solution is in excess in the mixture. The PI values of all these dispersions are very low and in the range of monodisperse distributions, respectively.

[1] B. Philipp, H. Dautzenberg, K.J. Linow, J. Koetz, W. Dawydoff, *Prog.Polym.Sci.* **1989**, *14*, 91.
[2] H. Dautzenberg, W. Jaeger, J. Koetz, B. Philipp, Ch. Seidel, D. Stscherbina, "*Polyelectrolytes: Formation, Characterisation and Application*", Carl Hanser Verlag, Munich 1994.
[3] E. Tsuchida, K. Abe, "*Interactions Between Macromolecules in Solution and Intermacromolecular Complexes*", Springer Verlag, Berlin 1981, Chap.1.
[4] H. Dautzenberg, J. Hartmann, S. Grunewald, F. Brand, *Ber. Bunsenges. Phys. Chem.* **1996**, *100*, 1024.
[5] V.A. Kabanov, A.B. Zezin, *Pure Appl. Chem.* **1984**, *56*, 343.
[6] T. Schindler, E. Nordmeier, *Polym. J.* **1994**, *26*, 1124
[7] C.K. Trinh, W. Schnabel, *Macromol. Chem. Phys.* **1997**, *198*, 1319.
[8] Y. Dan, Q. Wang, *Polym. Int.* **2000**, *49*, 551.
[9] V.A. Kabanov, A.B. Zezin, *Makromol. Chem.* **1984**, *6*, 259.
[10] E. Tsuchida, *J.M.S.- Pure Appl. Chem.* **1994**, *A31*, 1.
[11] N. Karibyants, H. Dautzenberg, H. Cölfen, *Macromolecules* **1997**, *30*, 7803.
[12] N.V. Pogodina, N.V. Tsvetkov, *Macromolecules* **1997**, *30*, 4897.
[13] D.V. Pergushov, H.M. Buchhammer, K. Lunkwitz, *Colloid Polym. Sci.* **1999**, *277*, 101.
[14] G. Petzold, A. Nebel, H.M. Buchhammer, K. Lunkwitz, *Colloid Polym. Sci.* **1998**, *276*, 125.
[15] H. M. Buchhammer, G. Petzold, K. Lunkwitz, *Langmuir* **1999**, *15*, 4306.
[16] H. M. Buchhammer, G. Petzold, K. Lunkwitz, *Colloid Polym. Sci.* **2000**, *278*, 841.
[17] M. Mende, G. Petzold, H. M. Buchhammer, *Colloid Polym. Sci.* **2002**, *280*, 342.
[18] H. Dautzenberg, E. Görnitz, W. Jaeger, *Macromol. Chem. Phys.* **1998**, *199*, 1561.
[19] W. M. Kulicke, M. Otto, A. Baar, *Makromol. Chem.* **1993**, *194*, 751.
[20] R. H. Müller, R. Schumann, „*Teilchengrößenmessung in der Laborpraxis*", Wiss-Verl.-Ges. mbH Stuttgart 1996, p. 25.

© 2004 WILEY-VCH Verlag GmbH & KGaA, Weinheim

Macromol. Symp. **2004**, *211*, 135-155

Aggregation Phenomena in Polyelectrolyte Multilayers Made from Polyelectrolytes Bearing Bulky Functional, Hydrophobic Fragments

André Laschewsky,[1,2] *Frank Mallwitz,*[1] *Jean-François Baussard,*[3] *Didier Cochin,*[3] *Peter Fischer,*[3] *Jean-Louis Habib Jiwan,*[3]*Erik Wischerhoff*[3]

[1]Universität Potsdam, Institut für Chemistry, Karl-Liebknecht-Straße 24-25, D-14476 Golm, Germany
E-mail: laschews@rz.uni-potsdam.de
[2]Fraunhofer-Institute for Applied Polymer Research FhG-IAP, Geiselberg-Straße 69, D-14476 Golm, Germany
[3]Université catholique de Louvain, Dept. of Chemistry, Place L. Pasteur 1, B-1348 Louvain-la-Neuve, Belgium
E-mail: habib@chim.ucl.ac.be

Summary: The functionalization of polyelectrolyte multilayers often implies the use of bulky functional fragments, attached to a standard polyelectrolyte matrix. Despite of the high density of non-charged, often hydrophobic substituents, regular film growth by sequential adsorption proceeds easily when an appropriate polyelectrolyte counter ion is chosen. However, the functional fragments may cluster or aggregate. This complication is particularly evident when using chromophores and fluorophores as bulky pendant groups. Attention has to be paid to this phenomenon for the design of functional polyelectrolyte films, as aggregation may modify crucially the properties. The use of charged spacer groups does not necessarily suppress the aggregation of functional side groups. Still, clustering and aggregation depend on the detailed system employed, and are not obligatory. In the case of cationic poly(acrylamide)s labeled with naphthalene and pyrene fluorophores, for instance, the polymers form intramolecular hydrophobic associates in solution, as indicated by strong excimer formation. But the polymers can undergo a conformational rearrangement upon adsorption so that they are decoiled in the adsorbed films. Analogous observations are made for polyanions bearing mesogenic biphenyls fragments. In contrast, polycations functionalized with the dye coumarin 343 show little aggregation in solution, but a marked aggregation in the ESA films.

Keywords: electrostatical self-assembly; functional polyelectrolytes; hydrophobic aggregation; layer-by-layer; polyelectrolyte multilayers

© 2004 WILEY-VCH Verlag GmbH & KGaA, Weinheim DOI: 10.1002/masy.200450710

Introduction

The layer-by-layer assembly of oppositely charged polyelectrolytes into ultrathin films, often named electrostatical self-assembly (ESA), has been well established in the past decade.[1-4] Developed only in the early 1990ies, many fundamental studies have been published since, and numerous potential applications have been proposed. A closer look to these developments shows that the vast majority of the work has been focused on a few standard polyelectrolytes only. Functionality for various uses has been mainly based on the sheer presence of the films (volume and barrier effects), on their surface modification, or on their patterning. Relatively little work has been dedicated yet to functional ESA films in which the function derives from molecular functionality of the constituting polyelectrolytes, i.e. to ESA films made of polyelectrolytes bearing functional groups. This is for sure partially due to the challenges inherent in the synthesis of complex polyelectrolytes. But moreover, the wide-spread opinion that a critical charge density is needed to prepare ESA films[5-8] seems to have discouraged the design of chemically functionalized systems: complex functional groups are in general rather bulky, rather hydro-phobic, and bear only few, or more often even no charged moieties. Fortunately, this opinion is not justified as exemplified in several studies. Rather than a critical charge density, the matching of the charge density of the polyelectrolyte pair is needed for successful ESA film growth.[9-11] Thus, appropriate selection of the polyelectrolyte *pairs* used allows the preparation of most different functional systems.[10-20] Still, the arrangement of the functional fragments in the films may pose problems, depending on the context. On the one hand, the functional fragments may be isotropically or anisotropically distributed within the films, and their alignment and orientation is difficult to control. On the other hand, the low polarity of the functional fragments, - compared to the ionic groups -, favours their aggregation and clustering. This is not only a matter of homogeneity or inhomogeneity on a given scale, but aggregation and clustering may drastically modify, or may even interfere with the desired functionality of chemical fragments.[18-24] This problematic is frequently overlooked.

We have therefore studied the use of several functionalized polyelectrolytes for ESA, bearing chromophores and fluorophores. Such functional groups may probe their own environment, thus providing information about possible aggregation phenomena. The standard and functional polyelectrolytes used are shown in Table 1.

© 2004 WILEY-VCH Verlag GmbH & KGaA, Weinheim

Table 1.   polycations (**1-9**) and polyanions (**10-13**) used

8 x=0.70 y=0.27 z=0.03   **9** x=0.86 y=0.08 z=0.06

© 2004 WILEY-VCH Verlag GmbH & KGaA, Weinheim

Experimental part

Materials

For all experiments, deionized water was purified by an Elgastat Maxima or a Millipore purification system (resistance 18.2 MΩ). Flash chromatography was performed on silica gel (Merck, 230-400 mesh). Branched poly(ethyleneimine) and poly(sodium 4-styrene-sulfonate) were purchased from Aldrich and used without further purification. Poly[diallyldimethyl-ammonium chloride] **1** was a gift from W. Jaeger (FhG-IAP, Golm). Poly[sulfonyl ethyl maleic acid monamide-*alt*-4'-cyanobiphen-4-yl-oxydodecyl vinyl ether] **13** was a gift of A. C. Nieuwkerk and A. T. M. Marcelis (Wageningen Agricultural University).[25] The synthesis of polymers **2**, **5-7**, and of poly[(sulfopropyl)methacrylate] was described previously.[26-28] The synthesis of polymers **8**, **9** and **12** will be described elsewhere.[29,30]

Synthesis of polycation **3**: Following a standard procedure,[31] 1.00 g (5.81 mmol) of 2-(naphth-1-yl)ethanol, 0.97 g (5.81 mmol) of 4-bromobutyric acid and 0.100 g of 4-dimethylaminopyridine in 30 ml of dry CH_2Cl_2 are cooled to 0 °C. A solution of 1.20 g (5.82 mmol) of dicyclohexyl-carbodiimide in 5 ml of dry CH_2Cl_2 is added drop wise. After cooling for another 5 h, the reaction mixture is stirred for 24 h at ambient temperature. The precipitated dicyclohexylurea is filtered off, and the solvent is evaporated. The crude 2-(naphth-1-yl)ethyl-4-bromobutanoate obtained is purified by flash-chromatography (eluent: CHCl$_3$) to yield 1.4 g (75%) of colourless powder. ^{1}H-NMR (200 MHz, CDCl$_3$): δ = 8.1 (m); 7.7.5-7.95 (m); 7.3-7.6 (m) [1H+2H+4H, CH aryl]; 4.46 (t) [2H, -CH$_2$O-]; 3.35-3.50 (m) [2H+2H,-CH$_2$Br, -CH$_2$-aryl]; 2.49 (t) [2H, -CH$_2$-COO-]; 2.13 (m) [2H, -C-CH$_2$-C-]. 0.19 g (0.59 mmol) of 2-(naphth-1-ylethyl)-4-bromobutanoate and 1.00 g (5.87 mmol) of poly[(3-N,N-dimethylaminopropyl)methacrylamide] are dissolved in a mixture of 10 ml of ethanol and 10 ml of nitromethane and stirred at 60 °C for 36 h under argon. Then, 0.79 ml of bromoethane (ρ=1.460, 10.6 mmol) are added and the mixture is stirred for 48 h at 40 °C. The resulting polymer **3** is precipitated into ethyl acetate, dissolved in ethanol and precipitated once more into ethyl acetate. Yield: 0.84 g (71%) of colourless hygroscopic powder.

Synthesis of polycation **4**: 0.25 ml (0.434 g; 3.47 mmol) of 2-bromoethanol, 1.00 g (3.47 mmol) of 4-pyren-1-ylbutyric acid and 0.04 g of 4-dimethylaminopyridine in a mixture of 20 ml of dry CH_2Cl_2 and 7 ml of dry THF are cooled to 0 °C. A solution of 0.716 g (3.47 mmol) of

© 2004 WILEY-VCH Verlag GmbH & KGaA, Weinheim

dicyclohexylcarbodiimide in 5 ml of dry CH_2Cl_2 is added drop wise. The cooling is maintained for 5 h and then the reaction mixture is stirred at ambient temperature for 24 h. After filtration of the precipitated dicyclohexylurea, the solvent is evaporated. The crude (2-bromoethyl)-4-pyren-1-yl-butanoate is purified by flash-chromatography (eluent: toluene), to yield 1.8 g (94%) of white solid. ^{1}H-NMR (200MHz, CDCl$_3$): δ = 7.7-8.35 (m) [9H, aryl]; 4.40 (t) [2H, -CH$_2$O-]; 3.35-3.55 (m) [4H,-CH$_2$Br, -CH$_2$-aryl]; 2.51 (t) [2H, -CH$_2$-COO-]; 2.21 (m) [2H, -C-CH$_2$-C-]. Polycation **4** is prepared from 1.00 g (5.87 mmol) of poly[(3-N,N-dimethylaminopropyl)-methacrylamide] and 0.23 g (0.59 mmol) of (2-bromoethyl)-4-pyren-1-ylbutanoate in analogy to polymer **3**. Yield: 0.92 g (75%) of colourless, hygroscopic solid.

Methods

NMR spectra were taken with a Gemini-200 spectrometer (Varian). UV/VIS spectra of the multilayer assemblies on quartz were recorded with a SLM-AMINCO DW-2000 spectrophotometer in the double beam mode, or with a spectrophotometer CARY 5E (Varian). A clean quartz plate served as reference. Ellipsometry was performed using polished Si-wafers as support with a Multiskop (Optrel GmbH, Berlin, Germany), assuming a refractive index of 1.5. Stationary fluorescence excitation and emission spectra were recorded with a 48000S SLM AMINCO spectrofluorometer and with a Fluorolog-3 spectrofluorometer Jobin Yvon – Spex system from Instruments S.A., Inc. (USA). A front-face configuration was used to record the spectra of the multilayer films. The spectra were corrected for the wavelength dependence of the detector. The X-ray reflectivity measurements were performed with a Siemens D5000 diffractometer for incidence angles lower than 4.5°, using a Cu-K$_α$ radiation and a secondary graphite monochromator. A good collimation of the beam was achieved by placing a knife-edge a few μm above the sample surface. The data were corrected for background scattering and variation of the illuminated area at very low angles of incidences. Data were analyzed by fitting a model consisting of a succession of thin sublayers on flat substrate using a matricial iteration formalism derived from Fresnel equations.[6]

Preparation of the polyelectrolyte multilayers

An automatic dipper (Kirstein & Riegler, Berlin, Germany) was used for the alternating polyelectrolyte deposition, working at room temperature in glass beakers without stirring.

© 2004 WILEY-VCH Verlag GmbH & KGaA, Weinheim

Multilayer films were built on fused quartz plates, or on polished Si-wafers. The polyions were dissolved in water with a concentration of 10^{-2} mol repeat unit per liter. Only in specified cases, HCl was added at a concentration of 10^{-2} M. The quartz plates used as substrates for the multilayers were cleaned as described.[32] Occasionally, initial buffer layers of cationic branched poly(ethyleneimine), anionic poly[(sulfopropyl)methacrylate] and cationic **7** were deposited onto the quartz substrates prior to the deposition of the systems studied. Multilayers were assembled by alternating dipping of the substrates into the solutions of the polycations and polyanions for 20 min each. Before changing the polyelectrolyte solutions, the substrates were rinsed by dipping into pure water three times for 1 min, to remove adhering excess solution. The resulting films made of x layer pairs, e.g. of polymers **2** and **10**, will be named $\{2/10\}_x$ in the following.

Results and Discussion

Polyelectrolyte multilayers can be easily grown from standard polyelectrolyte pairs with high charge density, for instance polymer films $\{1/10\}_x$, $\{1/11\}_x$, or $\{2/10\}_x$. Also, more complex polyelectrolytes, such as polyanion **12** bearing a crown ether can be easily deposited with polycation **1**, despite the high content of the uncharged crown fragment in the polymer (Figure 1). Following film growth by ellipsometry, regular, linear growth is observed (Figure 1a). However, thicker films tend to become turbid and the increase in optical density is no more linear (Figure 1b). Apart from other putative explanations, one may ask whether non-coulombic forces between the bulky and hydrophobic functional fragments may be responsible. In the UV-spectra, no change in the general spectral properties is observed so that there is no indication of special interactions. Still, one would not expect the simple chromophore to be sensitive to changes of its environment. But of course, a possible clustering of the crown ether unit might have important consequences for its complexation properties, for instance whether 1:1 or sandwiched 2:1 complexes would be formed with metal cations. Unfortunately, it is very difficult to derive in such systems information on the local arrangement of the functional groups, namely of the crown ethers in the films.

In order to obtain local structural information, - at least qualitatively -, we have looked into hydrophobically substituted polyelectrolytes, where the functional groups themselves can be used as probes, namely they are solvatochromic chromophores or fluorophores.

© 2004 WILEY-VCH Verlag GmbH & KGaA, Weinheim

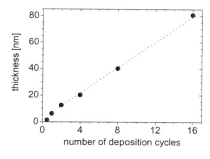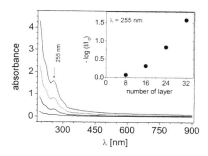

Figure 1.   Growth of multilayers $\{1/12\}_x$, as followed by (a) ellipsometry and by (b) UV-Vis spectroscopy: from bottom to top: 8, 16, 24 and 32 deposition cycles; inset: absorbance at $\lambda = 255$ nm as function of the number of deposition cycles x

A particularly interesting question arises when the modified polyelectrolytes that are employed for ESA films, are known to undergo hydrophobic aggregation in dilute aqueous solution. The general usefulness of some micellar polymers for the layer-by-layer deposition technique has been demonstrated before.[10, 26, 33-36] But it is not clear how the associative character in solution influences the structure of the adsorbed film, e.g. whether hydrophobic domains are preserved upon adsorption, as found in the case of bipolar amphiphiles,[37-40] or whether not.

The hydrophobically substituted polycations, **3** and **4** (see Table 1), can nicely illustrate the problem. Alike their unlabeled parent polymer **2**, both were prepared by quaternization from the same batch of precursor poly[(3-N,N-dimethylaminopropyl)methacrylamide], in order to have an identical degree of polymerization, thus avoiding eventual effects due to different molar masses. The ^{1}H NMR spectra of polymers **3** and **4** (Figure 2) show that the quaternization of the amino groups is not complete. Whereas the signal at 3.0 ppm is assigned to the protons of the methyl groups of the quaternized ammonium group, the signal at 2.85 ppm is assigned to the protons of the residual tertiary dimethylamine groups, which are mostly protonated in water. From the integration of the ^{1}H-NMR signals, the compositions of the tercopolymers were estimated. The ratio of the integrals of the signal of the aromatic protons and of the signals of the protons in the range of 3.6-2.5 ppm was used in the case of copolymer **3**. Similarly for copolymer **4**, the ratio of

© 2004 WILEY-VCH Verlag GmbH & KGaA, Weinheim

the integrals of signal of the aromatic protons and of the signals of the protons in the range of 3.5-2.6 ppm was used. The contents of hydrophobic substituents for copolymers **3** and **4** were accordingly estimated to be 20 mol-% naphthalene groups and 5 mol-% of pyrene groups, respectively.

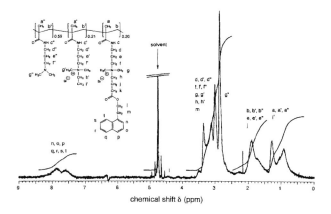

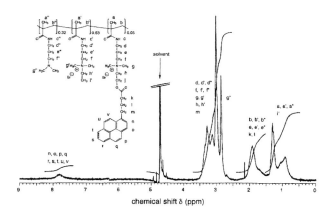

Figure 2.   ¹H-NMR spectra of polycation **3** (a) and **4** (b) in D₂O

© 2004 WILEY-VCH Verlag GmbH & KGaA, Weinheim

The choice of naphthalene and pyrene residues as hydrophobic groups offers the advantage that both are good fluorophores, so that they can simultaneously serve as probes. However, the residual tertiary amine groups due to the incomplete quaternization of the precursor polymer can quench the emission of the fluorophores. Though under ambient conditions (pH ≈ 5.8), the tertiary amines are expected to be mostly protonated in water, beforehand studies revealed still a notable increase in fluorescence intensity when reducing the pH value to 1 (Figure 3). This is attributed to the complete conversion of the residual amine groups into the hydrochloride form at low pH. Further addition of HCl increases the fluorescence intensities of the copolymers only slightly. Note that the shape of the emission spectra (see below) was not changed by the pH. The increase in fluorescence intensity is more pronounced for polymer **3** than for **4**, in agreement with the higher sensitivity of naphthalene to quenching by amines compared to pyrene.[41, 42]

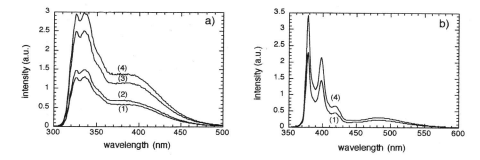

Figure 3.   Fluorescence emission spectra of polymers **3** (a) and **4** (b) in 10^{-2} M solution in
(1): water; (2): 10^{-3} M HCl ; (3): 10^{-2} M HCl ; (4): 10^{-1} M HCl .
Excitation at 290 nm for **3** or at 340 nm for **4**, respectively

Interestingly, the emission spectrum of copolymer **3** in aqueous solution shows not only the usual fluorescence band of the naphthyl group from 310 nm to about 400 nm with two maxima at 335 nm and 345 nm, but also an additional marked shoulder centered at about 390 nm (Figure 3a). The excitation spectra of the solution recorded at 335 nm, 420 nm and 440 nm correspond all to the absorption spectrum of polymer **3** in water. Therefore, the formation of a naphthalene dimer in ground state must be excluded. We assume the formation of a naphthalene

© 2004 WILEY-VCH Verlag GmbH & KGaA, Weinheim

excimer although such excimers are rare. But as the content of naphthyl groups in **3** is high, the formation of hydrophobic domains favouring excimer formation is most plausible.[43, 44] Analogously, the fluorescence emission spectrum of copolymer **4** in aqueous solution is characterized by the structured main emission between 350 nm and 450 nm, and an additional broad unstructured band centered at about 470 nm (Figure 3b). This band is assigned to the excimer of the pyrene,[45] suggesting hydrophobic association in water.

Polyanion **10** and the precursor polycation **2** are known to form layer-by-layer films of high quality.[26] Layer-by-layer assemblies of polyanion **10** with the functionalized analogs polycation **3**, or with **4** respectively, were first prepared from pure aqueous solutions, providing visually transparent films. The absorption band of **10** at 226 nm was used to monitor the regularity of the deposition process, because the absorbance of the naphthyl and pyrenyl groups were - though easily visible in the spectra - too low to quantify precisely the deposition process. Both systems $\{10/3\}_x$ and $\{10/4\}_x$ exhibit a linear increase of the absorbance with the number of deposition cycles (Figure 4), indicating a regular deposition process.

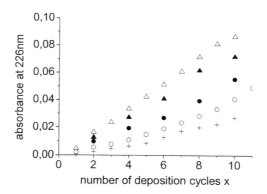

Figure 4.   Growth of layer-by-layer assemblies $\{10/3\}_x$ ($\triangle$, $\blacktriangle$) and $\{10/4\}_x$ ($\bigcirc$, $\bullet$), followed by the absorbance of **10** at 226 nm. Open symbols: films grown from aqueous solutions. Closed symbols: films grown from 10^{-2} M HCl solutions. Data for the films $\{10/2\}_x$ made from aqueous solutions are added for comparison (+)

© 2004 WILEY-VCH Verlag GmbH & KGaA, Weinheim

The amount of deposited polyanion **10** increases in the multilayer system $\{10/3\}_x > \{10/4\}_x > \{10/2\}_x$. Still, the density of charged groups in polycations **3** and **4** is close to the one in polycation **2**, and the degree of polymerization is the same. The thicker films in the case of the system $\{10/3\}_x$ may result from a notable contribution of hydrophobic interactions[26] or of favorable π-π interactions to layer growth, as the high number of naphthalene residues in **3** should well match with the phenyl groups of polyanion **10**. But though multilayer films can be grown regularly from pure aqueous solutions, the intensities of the fluorescence emission of these two systems $\{10/3\}_x$ and $\{10/4\}_x$ were very low with estimated quantum yields below of 1%, especially for the naphthalene based system. We attribute this finding to the high content of amino groups in the polymers which results in an efficient fluorescence quenching. Therefore, layer-by-layer assemblies were prepared from solutions of the cationic polymers in 10^{-2} M HCl, while the polyanion **10** was adsorbed from pure aqueous solutions as otherwise non-linear growth took place. In order to avoid any problem of intermediate storage of the films for higher deposition cycles (*cf.* ref. 26), a series of samples was prepared without any drying step. Again, the absorbance at 226 nm grows linearly with the number of dipping cycles, indicating the regularity of the process (Figure 4). As observed for films prepared from pure water, more **10** is adsorbed in $\{10/3\}_x$ films than in $\{10/4\}_x$ ones. However, the amount of polyanion **10** adsorbed varies when changing the pH. In the case of the system $\{10/3\}_x$, slightly less polymer is adsorbed when depositing from 10^{-2} M HCl, in the case of the system $\{10/4\}_x$, slightly more polymer is adsorbed. This somewhat surprising finding is attributed to two opposite effects which are superposed: the increase in ionic strength typically leads to thicker layers, whereas an increasing content of ionic groups, due to the increasing protonation of the residual amines with decreasing pH, results in thinner layers.[2] Apparently, the balance of these two effects is different for polycations **3** and **4**. In the case of polycation **3** with a high content of amines, the effect of increasing protonation seems to dominate, whereas in the case of polycation **4** with a low content of amines, the effect of increasing ionic strength seems to prevail.

The multilayers grown from 10^{-2} M HCl solutions show indeed improved fluorescence intensities, so that the environment of the hydrophobic groups may be probed. Most noteworthy, the emission spectra of the $\{10/3\}_x$ films show the absence of excimer emission in the naphthalene fluorescence (Figure 5). Similarly, no pyrene excimer emission is found for films $\{10/4\}_x$

© 2004 WILEY-VCH Verlag GmbH & KGaA, Weinheim

(Figure 6 and Figure 7). These systems are thus suited e.g. for Förster Resonance Energy Transfer experiments, taking advantage of the spectral match of the naphthyl and pyrenyl groups.[46] In fact, sequentially deposited films like **[10/4/10/3]**$_x$ and **[10/4/10/2/10/3/10/2]**$_x$ exhibit a strong pyrene emission at about 417 nm, compared to a low naphthalene emission in the region of 320 nm - 370 nm when excited at 290 nm. This is the wavelength where the absorbance by naphthalene chromophore compared to pyrene absorbance is maximized. For comparison, no pyrene fluorescence could be detected for the reference system **[10/4]**$_{10}$ when excited at 290 nm. Still, one has to keep in mind that quantum yields are low, so that it cannot rigorously be excluded that the fluorophores might not probe the average situation.

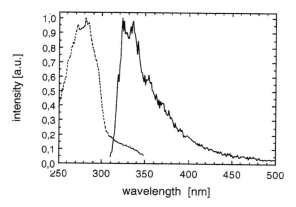

Figure 5.   Excitation (---) and emission (——) spectra of layer-by-layer assemblies {**10/3**}$_{10}$. Excitation and emission wavelengths are 290 nm and 372 nm, respectively

The fluorescence of these systems exhibits the peculiarity that shape of the emission spectra varies with time. Freshly prepared samples {**10/4**}$_x$ show a ratio of the intensity of the vibronic peaks I_1/I_5 of the emission band of about 1 (Figure 6). Samples stored for a few days exhibit a ratio of I_1/I_5 of about 1.5 (Figure 7). A closer look to the spectra shows that the ratio I_1/I_5 of freshly prepared films increases already slightly with the number of deposited layers (Figure 6). Although for substituted pyrenes, the ratio of I_1/I_5 is normally no more an accurate measure of the local polarity, it still provides a rough indication for it, thus suggesting a considerably more polar environment of the chromophores after annealing at room temperature.[47]

© 2004 WILEY-VCH Verlag GmbH & KGaA, Weinheim

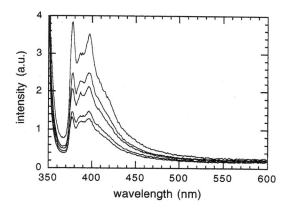

Figure 6.   Emission spectra of freshly prepared layer-by-layer assemblies {**10/4**}$_x$.
From bottom to top: x = 2, 4, 6, 8, 10 (Excitation at λ_{exc}= 345 nm)

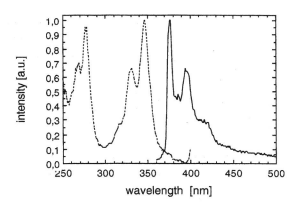

Figure 7.   Excitation (---) and emission (——) spectra of stored films {**10/4**}$_{10}$. Emission and excitation wavelengths are 417 nm and 347 nm, respectively

Two important conclusions impose from these spectral features, concerning the aggregation of the bulky hydrophobic fragments. First, the absence of excimer fluorescence suggests the disruption of the hydrophobic domains upon adsorption for both polymers **3** and **4**. This means that the polymers do not adsorb in the form of more or less deformed coils, let alone of micellar

© 2004 WILEY-VCH Verlag GmbH & KGaA, Weinheim

structures. In contrast, they seem to decoil efficiently when adsorbing on the support and to take an extended conformation. This interpretation is corroborated by X-ray reflectivity measurements performed on the polyelectrolyte films, providing thicknesses of less than 0.5 nm per polyanion-polycation layer pair, i.e. showing very thin adsorption layers. Second, the evolution of the ratio I_1/I_5 with storage indicates a continuous rearrangement of the polyelectrolyte complexes. Note that such a rearrangement is not reflected in the UV-absorbance spectra. There have been some other occasional reports about such rearrangements in the past, e.g. evidenced by solvatochromic shifts found for some colored polyelectrolytes,[10,26,48] or by X-ray reflectograms evolving with annealing.[33] The changing arrangement of the polymers in the films might be partially due to the more and more fading influence of the support in increasingly thicker films. Still, this does not explain the marked storage effect. Strangely, the pyrene labels seem to feel a rather hydrophobic environment in fresh films which becomes increasingly polar by time, finally approaching a value similar to the one found in the aqueous solution (*cf.* Figure 3b). We have no good explanation for this so far. In any case, the observed evolution of the I_1/I_5 ratio should exclude that the polymer rearrangement is related to micro phase separation of the hydrophobic fragments, as often found in pyrene labeled systems.[49, 50]

The behavior of the hydrophobically modified polycations **3** and **4** is not unique. In a similar way, the polyanion **13** is known to form micellar aggregates in aqueous solution.[25, 51-53] Hydrophobic association in this polymer is particularly strong due to the high density of hydrophobic groups, the long alkyl spacer present, and the mesogenic character of the cyanobiphenyl residue. Nevertheless, multilayers can be prepared from polyanion **13** with a number of polycations, for instance the standard polymer **5**. The evolution of the UV/Vis spectra with the number of deposition cycles shows a linear increase of the absorbance, demonstrating reproducible, regular film growth (Figure 8). A series of previous investigations showed that the hydrophobic association of this polymer can be studied by the spectral shift of the chromophore. In the isolated form, its absorbance maximum is at 297 nm, whereas in the aggregated form, a marked hypsochromic shift is observed.[51] In fact, strong chromophore aggregation typically results in marked spectral shifts, however, whether a hypsochromic or a bathochromic shift is observed, depends on the detailed structure of the aggregates and the resulting possibilities for interaction between the chromophores.[54]

© 2004 WILEY-VCH Verlag GmbH & KGaA, Weinheim

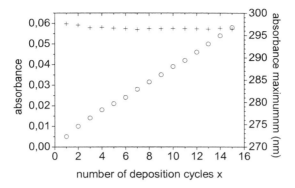

Figure 8.   Growth of layer-by-layer assemblies $\{5/13\}_x$ followed by UV/Vis spectroscopy. ($\bigcirc$) = absorbance, (+) = position of the absorbance maximum

In fact, aqueous solutions of polyanion **13** exhibit an absorbance band at $\lambda_{max}= 275$ nm,[25,53] indicating strong aggregation of the hydrophobic side chains. ESA multilayers $\{5/13\}_n$, however, have the absorbance band at $\lambda_{max}= 296$ nm (Figure 8). This indicates the dissociation of the chromophore aggregates, and the decoiling of the polyions upon adsorption. The position of the absorbance maximum is independent of the number of deposition cycles, except for the first two to three deposition cycles where a very small shift is found. Accordingly, no secondary aggregation etc. takes place in the multilayer films, as frequently observed.[10] This is noteworthy, as the functional fragments are not only strongly hydrophobic, but the 4-cyanobiphenyl moiety disposes also of a high dipole moment that would intuitively be expected to favor clustering. Though many examples of clustering of bulky, non-charged molecular fragments in ESA multilayers have been known, the here presented examples demonstrate, that clustering is not obligatory. Apparently, an appropriately chosen system may provide isolated functional fragments.

Having shown that it is a priori possible to avoid aggregation even of bulky hydrophobic groups, it would be desirable to have a rational strategy to do so on will (and not only by chance). Previous work has shown that inclusion complexation by cyclodextrins may be such a strategy.[24, 55] But this strategy is painstaking, and seems to be limited to particular molecular designs.[20] One may wonder whether the use of polyelectrolyte pairs that both dispose of bulky, but different

© 2004 WILEY-VCH Verlag GmbH & KGaA, Weinheim

hydrophobic substituents could minimize their aggregation in ESA films. The polarity difference between the molecular fragments in such systems is smaller, and thus the tendency for micro phase clustering and separation should be lower. In a preliminary study, we have therefore looked at the system $\{6/13\}_x$. Polyelectrolytes bearing azobenzene fragments, such as 6 or 7 seem to be particularly prone to aggregation in ESA films, independent whether they bear a charged moiety directly on the chromophore like 7, or not, like 6.[10] The ionene 6 shows an absorbance maximum at $\lambda_{max}= 458$ nm in methanol and ethanol, but at $\lambda_{max}= 488$ nm in aqueous solution. The markedly red-shifted value indicates chromophore association. As the strong absorbance of 6 at about 265 nm superposes the characteristic band of the cyanobiphenyl probe of 13 in such films, its spectral shift cannot be exploited. But the absorbance band in the visible of the azo-chromophore of 6 nevertheless is sufficiently conclusive. The absorbance maximum of films $\{6/13\}_x$ at about 478 nm (Table 2) demonstrates, that the use of two polyelectrolytes with hydrophobic fragments does not automatically prevent their aggregation. In fact, a decreasing content of charged groups in the polyanion seems to favour the red-shift of the band. From this point of view, the use of the overall less polar polyanion 13 poses no advantage compared to standard polyanions such as 10. Still, we note in this system that the absorbance maximum does not shift with increasing number of deposition cycles, suggesting a stable whatsoever arrangement of the chromophores in the films.

Table 2   Position of the absorbance maximum of polycation 6 in various multilayer films

polyanion used	absorbance maximum	reference
	nm	
poly(vinyl sulfate) 11	443	27
poly(3-sulfopropylitaconate)	443	27
poly(3-sulfopropylmethacrylate)	468	27
poly(styrene sulfonate) 10	478	27
13	478	this work
hectorite	498	this work
montmorillonite	500	27

© 2004 WILEY-VCH Verlag GmbH & KGaA, Weinheim

An alternative strategy to prevent clustering may be the use of ionic groups on, or close to, the functional groups: the repulsion of identically charged moieties should oppose to their clustering. As discussed above, for azobenzene moieties that are known for their tendency to aggregate,[56] this approach is not successful.[10, 57] But unfortunately, even for smaller chromophores that are less prone to aggregation, such as coumarins, this strategy does not work well. This can be exemplified by the use of the coumarin functionalized, analogous polycations **8** and **9** that are mainly distinguished by their different fluorophore content. Both polycations are suited for ESA multilayer films with various polyanions. The forms of their absorbance spectra are virtually identical. The chromophore exhibits an absorbance band with a maximum in the range of 440 nm - 450 nm in solution in organic solvents, whereas for the polycations in aqueous solutions, a shoulder in the absorbance band at about 435 nm appears, suggesting some aggregation of the fluorophore for both polymers (Figure 9), similar to the behavior of polyions **3**, **4** and **13**.

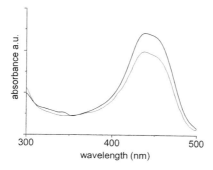

Figure 9.   UV/Vis spectra of coumarin functionalized polycations **8** (top curve) and **9** (bottom curve) in aqueous solution

But different from their cases discussed above, the aggregates are not disrupted upon polymer adsorption. In the contrary, for polycation **8** with only 3 mol-% of dye, the absorbance band shifts to about 415 nm in multilayers {**8**/**10**}$_x$, pointing to modified aggregation in the films. The extent of the spectral shift suggests the presence of aggregates of limited size (Figure 10a). Alternatively, the spectral shift might be attributed to a solvatochromic effect, though the value seems very large for the coumarin dye. The situation differs considerably for polycation **9** with the higher dye content. The complex spectrum, consisting of a broad band between 400 and 480 nm,

© 2004 WILEY-VCH Verlag GmbH & KGaA, Weinheim

and a strong new band at about 370 nm, indicates that the arrangement of the chromophores is not uniform in multilayers $\{9/10\}_x$. The major absorbance band at about 370 nm shows an extremely strong hypsochromic shift compared to the band of the isolated chromophore, pointing to the formation of large H-aggregates (Figure 10, curves 2 and 3).[29] The aggregation of the dye can be influenced by the choice of the solvent (Figure 10b). In fact, solvent effects were also reported for the aggregation in films of bola amphiphiles bearing oligo(thiophene)s.[40] For instance, deposition of **9** from water/DMF mixtures favors aggregate formation, as indicated by the strong hypsochromic shift of the absorbance band from about 450 nm to 375 nm. In contrast, deposition from water/isopropanol mixtures provides a spectrum similar to the one of the isolated chromophore. Accordingly, dye aggregation is efficiently suppressed. The differences found for the multilayers, however, are not reflected in the solution spectra: both mixed solvents provide spectra with the band at about 450 nm, i.e. the chromophores are isolated.

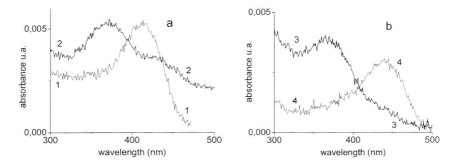

Figure 10.   UV/Vis spectra of coumarin functionalized multilayer films.

a): (1) = $\{8/10\}_5$, (2) = $\{9/10\}_5$.

b): films $\{9/10\}_5$, (3) = deposited from solutions in DMF/water 1v/1v, (4) = deposited from solutions in isopropanol/water 1v/1v

This means that in one or the other way, adsorption and/or drying kinetics play a role for the aggregation of functional fragments. So far, we cannot see an intelligible relationship behind the case by case observations. The only clear effect comes form the content of functional fragments, as polycations **8** and **9** exemplify: not surprisingly, as higher the content is, the higher is the tendency for aggregation.

© 2004 WILEY-VCH Verlag GmbH & KGaA, Weinheim

Conclusions

Polyelectrolyte films can be prepared from polyelectrolytes which are heavily substituted with large functional, hydrophobic fragments without particular problems. A priori, there is no need for a high charge density in the polymers. Though a strong tendency is found for many systems to form clusters and aggregates of the functional groups, it is possible to keep individual fragments isolated. This is also true for polyions with rather high contents of non-ionic fragments. In fact, experimental findings even strongly suggest that hydrophobic aggregates that are present in the solutions used for the adsorption process, can dissociate upon adsorption. The observations imply conformational transitions with complete decoiling of the polymers upon adsorption. Noteworthy, a rearrangement of chromophore fragments can still take place in the deposited films upon storage. The presence or absence of aggregates is a sensitive function of the solvent from which the polymers are adsorbed. At present, it is not possible to establish clear guidelines and rules what to do to avoid clustering/aggregation phenomena (except for dilution), - or to exploit them, respectively. Nevertheless, in the strong thrive to employ ESA films for various potential applications, one should be aware of the possible complications and pitfalls resulting from clustering or aggregation of functional molecular fragments in the films.

Acknowledgment

The authors thank A. Jonas and A. Moussa (U.C. Louvain, Dept. of Materials Science and Engineering) for assistance for X-Ray facilities and helpful discussions, W. Jaeger (FhG-IAP) for the gift of poly[diallyldimethylammonium chloride], H.-J. Hold and W. Mickler for the gift of monomeric crown ether, A. C. Nieuwkerk and A. T. M. Marcelis (Wageningen Agricultural University) for the gift of polyanion 13, and M. Van der Auweraer and E. Rousseau (K.U. Leuven) for help with some fluorescence studies. Financial support for the work was provided by the Fonds National de la Recherche Scientifique of Belgium, by the DG Recherche Scientifique of the French Community of Belgium (convention ARC 00/05-261), and by the Fraunhofer society.

© 2004 WILEY-VCH Verlag GmbH & KGaA, Weinheim

[1] G. Decher, M. Eckle, J. Schmitt, B. Struth, *Curr. Opin. Colloid Interface Sci.* **1998**, *3*, 32.
[2] P. Bertrand, A. Jonas, A. Laschewsky, R. Legras, *Macromol. Rapid Commun,* **2000**, *21*, 319.
[3] G. Decher, J. B. Schlenoff, Eds., *"Multilayer Thin Films. Sequential Assembly of Nanocomposite Materials"*, Wiley-VCH Publishers, Weinheim 2003.
[4] S. K. Tripathy, J. Kumar, H. S. Nalwa, Eds., *"Handbook of Polyelectrolytes and their Applications 1. Polyelectrolyte-Based Multilayers, Self-Assemblies and Nanostructures"*, American Scientific Publishers, Los Angeles 2002.
[5] N. G. Hoogeveen, M. A. Cohen Stuart, G. J. Fleer, M. R. Böhmer, *Langmuir* **1996**, *12*, 3675.
[6] X. Arys, A. M. Jonas, B. Laguitton, A. Laschewsky, R. Legras, E. Wischerhoff, *Prog. Org. Coatings* **1998**, *43*, 108.
[7] S. T. Dubas, J. B. Schlenoff, *Macromolecules*, **2001**, *34*, 3736.
[8] B. Schoeler, G. Kumaraswamy, F. Caruso, *Macromolecules* **2002**, *35*, 889.
[9] T. Okubo, M. Suda, *Colloid Polym. Sci.* **2002**, *280*, 533.
[10] P. Fischer, A. Laschewsky, E. Wischerhoff, X. Arys, A. Jonas, R. Legras, *Macromol. Symp.* **1999**, *137*, 1.
[11] M. Koetse, A. Laschewsky, A. M. Jonas, W. Wagenknecht, *Langmuir* **2002**, *18*, 1655.
[12] D. Cochin, M. Paßmann, G. Wilbert, R. Zentel, E. Wischerhoff, A. Laschewsky, *Macromolecules* **1997**, *30*, 4775
[13] G. Zotti, S. Zecchin, A. Berlin, G. Schiavon, G. Giro, *Chem. Mater.* **2001**, *13*, 43.
[14] M. Paßmann, R. Zentel, *Macromol. Chem. Phys.* **2002**, *203*, 363.
[15] A. Ziegler, J. Stumpe, A. Toutianouch, B. Tieke, *Coll. Surf. A* **2002**, *A198-200*, 777.
[16] D. G. Kurth, M. Schütte, J. Wen, *Coll. Surf. A* **2002**, *A198-200*, 633.
[17] J. Halfyard, J. Galloro, M. Ginzburg, Z. Wang, N. Coombs, I. Manners, G. A. Ozin, *Chem. Commun.* **2002**, 1746
[18] T. S. Lee, J. Kim, J. Kumar, S. Tripathy, *Macromol. Chem. Phys.* **1998** *199*, 1445.
[19] G. A. Lindsay, M. J. Roberts, A. P. Chafin, R. A. Hollins, L. H. Merwin, J. D. Stenger-Smith, R. Y. Yee, P. Zarras, *Chem. Mater.* **1999**, *11*, 924.
[20] M. Koetse, A. Laschewsky, A. M. Jonas, T. Verbiest, *Coll. Surf. A* **2002**, *A198-200*, 275.
[21] F. Saremi, G. Lange, B. Tieke, *Adv. Mater.* **1996** *8*, 923.
[22] K. Araki, M. J. Wagner, M. S. Wrighton, *Langmuir* **1996** *12*, 5393.
[23] N. Kometani, H. Nakajima, K. Asami, Y. Yonezawa, O. Kajimoto, *J. Phys. Chem. B.* **2000**, *104*, 9630.
[24] P. Fischer, M. Koetse, A. Laschewsky, E. Wischerhoff, L. Jullien, T. Verbiest, A. Persoons, *Macromolecules* **2000**, *33*, 9471.
[25] A. C. Nieuwkerk, E. J. M. Van Kan, A. Koudijs, A. T. M. Marcelis, E. J. R. Sudhoelter, *Eur. J. Org. Chem.* **1999**, *1*, 305.
[26] D. Cochin, A. Laschewsky, *Macromol. Chem. Phys.* **1999**, *200*, 1.
[27] A. Laschewsky, E. Wischerhoff, M. Kauranen, A. Persoons, *Macromolecules* **1997**, *30*, 8304.
[28] D. Cochin, P. Hendlinger, A. Laschewsky, *Colloid Polymer Sci.* **1995**, *273*, 1138.
[29] J. F. Baussard, J. L. Habib-Jiwan, A. Laschewsky, submitted
[30] H.-J. Hold, A. Laschewsky, F. Mallwitz, W. Mickler, to be published
[31] A. Neises, W. Steglich, *Angew. Chem.* **1978** , *90*, 556.
[32] P. Y. Vuillaume, A. M. Jonas, A. Laschewsky, *Macromolecules* **2002**, *35*, 5004.
[33] X. Arys, A. Laschewsky, A.M. Jonas, *Macromolecules* **2001**, *34*, 3318.
[34] K. Glinel, A. M. Jonas, A. Laschewsky, *Macromolecules* **2001**, *34*, 5267.
[35] Z. Dai, H. Möhwald, *Chem. Eur. J.* **2002**, *8*, 4751.
[36] D. W. Kim, A. Blumstein, J. Kumar, L.A. Samuelson, B. Kang, C. Sung, *Chem. Mater.* **2002**, *14*, 3925.
[37] G. Decher, J. D. Hong, *Ber. Bunsenges. Phys. Chem.* **1991**, *95*, 1430.
[38] X. Zhang, M. Gao, X. Kong, Y. Sun, J. Shen, *Chem. Commun.* **1994**, 1055.
[39] F. Saremi, B. Tieke, *Adv. Mater.,* **1995**, *7*, 378.
[40] J. Locklin, J. H Youk, C. J. Xia, M. K. Park, X. W. Fan, R. C. Advincula, *Langmuir* **2002**, *18*, 877.
[41] F. Meeus, M. Van der Auweraer, F. C. De Schryver, *J. Am. Chem. Soc.* **1980**, *102*, 4017.
[42] M. A. Winnik, S. M. Bystryak, Zh. Liu, J. Siddiqui, *Macromolecules* **1989**, *22*, 734.
[43] D. M. Gravett, J. E. Guillet, *Macromolecules* **1995**, *28*, 274.
[44] J. E. Guillet, W. A. Rendall, *Macromolecules* **1986**, *19*, 224.
[45] F. M. Winnik, *Chem. Rev.* **1993**, *93*, 587.
[46] J. R. Lakowicz, *"Principles of Fluorescence Spectroscopy"*, 2nd ed., Kluwer Academic/Plenum publishers, New York 1999, p.368.

© 2004 WILEY-VCH Verlag GmbH & KGaA, Weinheim

[47] K. Kalyanasundaram, J. K. Thomas, *J. Am. Chem. Soc.* **1977**, *99*, 2039.

[48] J. L. Casson, H. L. Wang, J. B. Roberts, A. N. Parikh, J. M. Robinson, M. S. Johal, *J. Phys. Chem. B* **2002**, *B106*, 1697.

[49] C. H. Lochmüller, A. S. Colborn, M. L. Hunnicut, J. M. Harris, *J. Am. Chem. Soc.* **1984**, *106*, 4077.

[50] I. Yamazaki, N. Tamai, T. Yamazaki, *J. Phys. Chem.* **1987**, *91*, 3572.

[51] A. C. Nieuwkerk, A. T. M. Marcelis, E. J. R. Sudhoelter, *Macromolecules* **1995**, *28*, 622.

[52] A. C. Nieuwkerk, A. T. M. Marcelis, E. J. R. Sudhoelter, *Langmuir* **1997**, *13*, 3325.

[53] A. C. Nieuwkerk, Ph.D. thesis, Wageningen Agricultural University (NL) 1998.

[54] M. Kasha in: "*Spectroscopy of the Excited State*", B. D. Bartolo Ed., p.337, Plenum Press, New York 1976.

[55] M. Dreja, I. T. Kim, Y. Yin, Y. Xia, *J. Mater. Chem.* **2000**, *10*, 603.

[56] T. Kunitake, *Angew. Chem.* **1992**, *104*, 692.

[57] M. K. Park, R. C. Advincula, *Langmuir* **2002**, *18*, 4532.

© 2004 WILEY-VCH Verlag GmbH & KGaA, Weinheim

Self-Assembly in Ternary Systems: Cross-Linked Polyelectrolyte, Linear Polyelectrolyte and Surfactant

Alexander B. Zezin,* Valentina B. Rogacheva, Olga A. Novoskoltseva, Victor A. Kabanov

Faculty of Chemistry, Polymer Department, Lomonosov Moscow State University, Lenin Hills, Moscow 119992, Russia
E-mail: zezin@genebee.msu.su

Summary: The competitive interactions in ternary systems consisting of a slightly cross-linked polyelectrolyte hydrogel and the mixture of linear polyelectrolyte and micelle forming surfactant both oppositely charged relative to the polyelectrolyte network were studied. It was shown that the equilibrium in the competitive reactions depends on the linear polyion charge density and the length of the surfactant aliphatic radical. Dependently on these characteristics the interpolyelectrolyte complex formed by cross-linked and linear polyelectrolytes can uptake surfactant ions from water solution transforming into the cross-linked polyelectrolyte-surfactant complex and releasing the linear polyelectrolyte or vice versa. The ternary systems of this kind are perspective to design the novel family of delivery constructs.

Keywords: competitive sorption; hydrogels; polyelectrolyte; surfactants; self-assembly

Introduction

Sorption of linear polyions,[1-3] proteins[4,5] and micelle forming ionic surfactants[6-11] by oppositely charged slightly cross-linked highly swollen polyelectrolyte hydrogels from water solutions is driven by cooperative electrostatic binding of sorbate species to polyionic fragments of the gel network. As a result cross-linked interpolyelectrolyte complex (#IPEC) or polyelectrolyte-surfactant complex (#PESC) are formed in the gel phase decreasing its volume by more than two orders of magnitude. By and large the process represents a heterogeneous reaction propagating inward the peace of gel. Its mechanism was established in our early studies.[1, 3, 6, 10] At the intermediate stage of such a reaction if it proceeds in the absence of an external salt, the sample remains heterogeneous consisting of a newly formed #IPEC or #PESC shell separated by the sharp boundary from the unreacted original gel core. The thickness of the shell can be easily controlled by variation of experimental conditions.[3, 5] It opens the perspective to create novel techniques for constructing either polycomplex layers or multilayers on the hydrogel surface. Moreover, hydrogels loaded with linear polyions, proteins or ionic amphiphils are able to release these constituents at certain conditions into surroundings in a controllable manner. The latter can be used to design novel function-specific complex constructs, in particular for controlled drug release. In contrast to a number of earlier studied binary hydrogel-polyelectrolyte or hydrogel-surfactant systems the ternary systems in which a cross-linked polyelectrolyte hydrogel (#PE) interacts either with linear polyelectrolyte (*l*PE) or an ionic surfactant (Sf) at the same time are still in the opening stage of research.[12-14] A series of such systems is the subject of the present study.

Experimental Part

Materials

Acrylic acid (AA) and N,N-dimethylaminoethylmethacrylate (DMAEM) were purified by vacuum distillation. N,N-dimethyl-N-ethylaminoethylmethacrylate bromide (DMEAEMB) was synthesized by alkylation of DMAEM with ethyl bromide via the dropwise addition of DMAEM to 5-fold excess of ethyl bromide on stirring and ice-cooling. DMEAEMB formed as a white precipitate was washed by ethyl ether and dried in vacuum.

The cross-linked polyelectrolytes: poly(acrylic acid) (#PAA), poly(N,N-dimethylaminoethylmethacrylate) (#PDMAEM) and poly(N,N-dimethyl-N-ethylaminoethyl-methacrylate bromide) (#PDMEAEMB) were synthesized by free radical copolymerization of AA with DMAEM in 10 wt.-% and AA with DMEAEMB in 20 wt.-% aqueous solutions with 1 mol-% of N,N-methylenebisacrylamide as a cross-linker. Polymerization was initiated by ammonium persulfate/sodium methabisulfite redox system (0.2% of monomer mass) and performed at 40 °C for 24 hours under argon atmosphere.[6, 10] #PDMAEM was prepared in salt form (#PDMAEM·HCl) by adding equivalent amount of HCl to the reaction mixture.

© 2004 WILEY-VCH Verlag GmbH & KGaA, Weinheim

After polymerization was completed the obtained hydrogels were immersed in a large amount of distilled water to wash out the residual chemicals. The wash water was repeatedly changed every 2-3 days during a month. #PAA hydrogel was transformed into cross-linked poly(sodium acrylate) (#PANa) by complete neutralisation with NaOH (to pH=9). The degree of swelling of equilibrium-water-swollen gels, $H = \dfrac{m_{sw} - m_{dry}}{m_{dry}}$ (m_{sw} and m_{dry} are the masses of equilibrium swollen and dry samples, correspondingly), was about 600-1000 for different gels.

Linear polyelectrolytes: poly(N-ethyl-4-vinylpyridinium bromide) (PEVPB) of weight-average degree of polymerization $\bar{P}_w = 700$ ($M_w/M_n = 1.2$), poly(N,N-dimethylaminoethylmethacrylate) (PDMAEM), $\bar{P}_w = 640$, poly(N,N-diallyl-N,N-dimethylammonium chloride) (PDADMAC), $\bar{P}_w = 3000$, poly(sodium acrylate) (PANa), $\bar{P}_w = 200$, and polyethylenimine (PEI), $\bar{P}_w = 1400$, were purchased from "Tokio Kasei" (Japan). Luminescently tagged poly(methacrylic acid) (PMA^t) was prepared by the reaction of pyrenyldiazomethane with poly(methacrylic acid), $\bar{P}_w = 1000$ [15].

The cationic surfactants: dodecylpyridinium chloride (DDPyC), cetylpyridinium chloride (CPyC), dodecyltrimethylammonuim bromide (DDTMAB), cetyltrimethylammonuim bromide (CTMAB), cetylamine (CA) and anionic surfactants: caprylic acid ($C_7H_{15}COOH$), caprynic acid ($C_9H_{19}COOH$), sodium dodecylsulfate (SDS) and sodium dodecylbenzosulfonate (SDBS) were purchased from "Serva". Each of them was purified by repeated recrystallization from of ethanol/acetone (15:85 v/v) mixed solvent before use. Caprilic and caprinic acids were transformed into sodium salts by neutralization with NaOH.

Stoichiometric (charge ratio 1:1) #IPECs and #PESCs were prepared at room temperature by immersing equilibrium-swollen hydrogel samples with mass of 2-5g into the aqueous solutions containing two-fold excess of oppositely charged /PE or respectively Sf [1,3,6,10]. After a period of time required for completing the interpolyelectrolyte or polyelectrolyte-surfactant reaction, the #IPECs containing equimolar amounts of #PE and /PE oppositely charged monomer repeating units or correspondingly #PESCs including equimolar amounts of #PE and Sf oppositely charged ionic fragments were formed. As a result transparent piece of

© 2004 WILEY-VCH Verlag GmbH & KGaA, Weinheim

the initial highly swollen hydrogel was transformed into a slightly swollen opaque #IPEC or #PESC peace with mass of 20 - 50 mg. The reactions were accompanied by a decrease in the volume of a reacted gel by more than 2 orders of magnitude.

Measurements

Concentrations of the reagents absorbing UV-light were measured spectrophotometrically in water solutions: DDPyC and CPyC at $\lambda = 259$ nm ($\varepsilon = 4\ 100$), SDBS at $\lambda = 260$ nm ($\varepsilon = 420$), PEVPB at $\lambda = 257$ nm ($\varepsilon = 2\ 700$), PMA^t at $\lambda = 342$ nm ($\varepsilon = 50\ 000$), PEI in form of tetra-coordinating complex with Cu^{2+} at $\lambda = 630$ nm ($\varepsilon = 200$).[16] Concentration of PDADMAC was estimated by turbidimetric titration with linear poly(sodium styrene sulfonate).[17] Spectrophotometric measurements were carried out by "Hitachi 150-20" (Japan) UV/VIS spectrophotometer.

Concentration of alkylcarboxylates and PANa was measured by potentiometric titration with HCl using "Radiometer pHM-83" (Denmark) pH-meter. The measurement accuracy was ± 0.02 pH units. Concentration of SDS was determined either by potentiometric titration of SDS mixture with two-fold excess of linear PDMAEM or using surfactant-selective electrode as described in.[12]

X-ray measurements were performed with an automatic diffractometer equipped with a two-dimensional high-resolution detector (Institute of Crystallography, Russian Academy of Sciences) using monochromatic (1.54 Å) CuK_α radiation. The averaging over diffraction cones[18] was used to obtain the dependence of intensity on the scattering angle, θ. Bragg parameters, d, were calculated from measured scattering vectors $Q = (4\pi/\lambda) \times \sin\theta$ where $\lambda = 1{,}54$ Å is the wavelength of the incident beam using the common equation $d = 2\pi/Q$.

Results and Discussion

Generally speaking in a ternary system consisting of #PE immersed in water solution *l*PE and Sf both oppositely charged relative to the #PE, either #IPEC or #PESC can be formed. In other words *l*PE and Sf are in competition for binding to #PE. The systems under study include either the cationic hydrogels (#PDMEAEMB, #PDMAEM·HCl) combined with the anionic *l*PE and Sfs, or the anionic hydrogel (#PANa) combined with the cationic *l*PE and Sfs. To study the competitive interactions in ternary systems the #IPEC sample originally prepared

© 2004 WILEY-VCH Verlag GmbH & KGaA, Weinheim

via reaction between #PE and oppositely charged /PE as described in the experimental part was placed into the aqueous solution of the Sf. In the preset time intervals the concentration of Sf and /PE in the environmental solution as well as the mass of the sample were measured.

It was found that some #IPECs sorb Sf ions charged similarly to the /PE incorporated into the original #IPEC if Sf concentration, C_{Sf}, was sufficiently high.[14] Sorption of Sf is accompanied by the release of the originally complexed /PE into surroundings and temporal changes in mass of the sample. The experimental data obtained are presented in Table 1 as ultimate extent of Sf sorption $(N_{Sf}/N_{\#PE})_s$ or /PE release $(N_{/PE}/N_{\#PE})_r$. N_{Sf} is the maximal mol amount of sorbed Sf estimated by the decrease in its concentration in the solution, $N_{\#PE}$ is the total base mol amount of #PE calculated as $N_{\#PE} = \dfrac{m_{sw}}{M_o \cdot (H+1)}$, where M_o is the molecular mass of the #PE monomer unit, and $N_{/PE}$ is the maximal base mol amount of /PE released determined by the increase in its concentration in the solution. In these experiments initial Sf concentrations were of 2 c.m.c. while initial Sf mol amounts were twice as much as initial base mol amounts of #PE.

Table 1. The data on the Sf sorption by #IPEC and /PE release from #IPEC.

№	SYSTEM	$(N_{Sf}/N_{\#PE})_s$	$(N_{/PE}/N_{\#PE})_r$
1	#IPEC (#PDMEAEM-PMA^t) + sodium caprinate	1.0	1.0
2	#IPEC (#PDMEAEM-PA) + sodium caprilate	1.0	1.0*
3	#IPEC (#PDMEAEM-PA) + sodium caprinate	0.9	1.0*
4	#IPEC (#PDMEAEM-PA) + SDS	0.9	1.0*
5	#IPEC (#PDMEAEM-PA) + SDBS	1.0	1.0*
6	#IPEC (#PA-PEVP) + CA·HCl	1.0*	0.9
7	#IPEC (#PA-PEVP) + CTMAB		0.9
8	#IPEC (#PA-PDADMA) + CPyC	1.0	1.0*

System 1 demonstrates clearly that the processes of Sf sorption and /PE release are correlated: the sample of #IPEC (#PDMEAEM-PMA^t) sorbs equimolar amount of sodium caprinate with respect to the base mol amounts of #PE and as a result eliminates the same mol amount of

© 2004 WILEY-VCH Verlag GmbH & KGaA, Weinheim

PMA^t into the surroundings. These data were obtained in the course of continuous measuring of the concentrations of both Sf (by means of potentiometric titration) and tagged PMA^t (by means of UV-spectrophotometry).

The results obtained are evident of quantitative replacement of Sf ions by IPE. In other words competitive sorption of Sf ions can be represented as transformation of stoichiometric #IPEC into stoichiometric #PESC as it is sketched by Reaction (1) (simple counter ions are omitted):

$$\text{#IPEC (#PE-}I\text{PE) + Sf} \rightleftharpoons \text{#PESC (#PE-Sf)} + I\text{PE} \qquad (1)$$

Note, that in other systems listed in Table 1 either the decrease in Sf concentration (lines 2-5, 8) or the increase in IPE concentration (lines 6,7) in the solutions was determined experimentally with sufficiently high accuracy. To make sure, that the Sf uptake is accompanied by the IPE release and the equilibrium in Reaction (1) is completely shifted to the right we have studied whether the reversed process takes place. In this case the #PESC samples obtained in advance as described in the experimental part and our previous studies[6,10] were put into aqueous solutions of the corresponding IPE. During long time (a few months) we could not find any decrease in IPE concentration (systems 2-5) or Sf release (system 6, 8) using UV-spectrophotometry and potentiometric tittration as noted in the experimental part. These data are presented in Table 1 as values labelled by asterisk.

The above unambiguously results reveal that actually in the systems listed in Table 1 the Sfs are able to substitute completely IPE incorporated in #IPEC. Hand in hand with these systems there are those in which Reaction (1) is reversed, i.e. the #IPEC samples do not sorb Sf ions from aqueous solutions, on the contrary, the samples of the corresponding stoichiometric #PESC obtained in advance[6, 10] uptake IPE from aqueous media releasing constituent Sf into the surroundings. The experimental data expressed in terms of ultimate extent of IPE sorption $(N_{I\text{PE}}/N_{\text{#PE}})_s$ or Sf release $(N_{\text{Sf}}/N_{\text{#PE}})_r$ are listed in Table 2. Here we used the same procedures as those for the analyses of the systems shown in Table 1.

Table 2. The data on the IPE sorption by #PESC and Sf release from #PESC.

№	SYSTEM	$(N_{I\text{PE}}/N_{\text{#PE}})_s$	$(N_{\text{Sf}}/N_{\text{#PE}})_r$
1	#PESC (#PA-DDPy)+PEI·HCl	1.0*	0.9
2	#PESC (#PA-CPy)+PEI·HCl	1.0*	1.0
3	#PESC (#PA-DDTMA)+PEVPB	1.0	

© 2004 WILEY-VCH Verlag GmbH & KGaA, Weinheim

The data presented in Table 2 demonstrate actually that in these systems the equilibrium in Reaction (1) is shifted from right to left, i.e. toward formation of the corresponding #IPECs. Generally the question of the factors controlling the above forward and reverse competitive reactions and position of equilibrium described by Equation (1) is of a crucial importance. To gain a better understanding of the problem we obtained experimentally sorption isotherms for different ternary systems. Some of them are represented in Figure 1 as the dependence of the sorption extent, $F=(N_{Sf}/N_{\#PE})_s$, versus the logarithm of the equilibrium Sf concentration in a solution. The points used to construct the curves were obtained for #IPECs incubated in aqueous solutions of Sf for a month. The amount of Sf sorbed was calculated as the difference between the initial and equilibrium Sf concentration in the solution surrounding #IPEC sample.

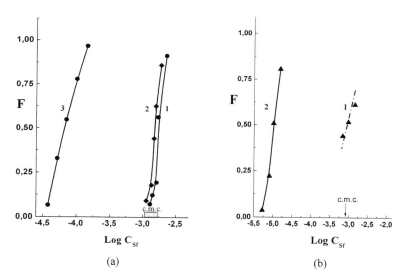

(a)   (b)

Figure 1. (a) - Sorption isotherms of SDBS by stoichiometric #IPEC (#PDMEAEM-PA) at 22 °C (1), 45 °C (2) and by #PDMAEM·HCl hydrogel at 20 °C (3);[10] mass of the initial #IPEC sample 0.05 g, mass of the initial hydrogel is 5g.
(b) - Sorption isotherms of CPy cations by stoichiometric #IPEC (#PA-PDADMA) at 22 °C (1), and by #PANa hydrogel at 40 °C (2);[10] mass of the initial #IPEC samples 0.02 g, mass of the initial gel is 2 g.

© 2004 WILEY-VCH Verlag GmbH & KGaA, Weinheim

Figure 1a shows that sorption of DBS anions by #IPEC (#PDMEAEM-PA) either at 22 °C (curve 1) or at 45 °C (curve 2) starts at SDBS concentrations located in its c.m.c. region (in literature $(1.2-1.7) 10^{-3}$ at 30 °C and $1.7 10^{-3}$ mol/l at 55 °C[19]). The same is true for sorption of CPy cations (c.m.c. is $9.0 10^{-4}$ mol/l at 25 °C[20]) by #IPEC (#PA-PDADMA) (Figure 1b, curve 1). It is significant that in contrast to the #IPECs original #PDMEAEM HCl or #PANa hydrogels start to sorb SDBS (Figure 1a, curve 3) or CPy cations (Figure 1b, curve 2) at Sf concentrations two orders of magnitude below the corresponding c.m.c. values.

The above difference finds a simple qualitative explanation, which involves taking account of translational entropy of low molecular ions in the system. The mere fact that micellization of ionic Sf unimers is induced by oppositely charged linear and cross-linked polyelectrolytes at critical aggregation concentrations much lower than c.m.c. is well known[6, 8, 9, 10, 21] and treated theoretically.[9] Micellization of free ionic Sf is driven by hydrophobic interaction of hydrocarbon tails. At the same time it requires immobilisation of simple counter ions in the vicinity of a micelle surface to neutralize the charge of condensed Sf ionic heads. The latter provides a certain entropy loss. The situation drastically changes if Sf ions are condensed on PE chain so that simple counter ions are replaced and released into environmental solution as it is shown in Scheme 1a. In such case the above entropy loss is nearly prevented. It gives a certain relative gain in free energy, which provides micellization of Sf ions at much lower Sf concentration. Nothing of that kind may happen if ionic Sf interacts with preformed IPEC or #IPEC consisting of high linear charged density polyions.

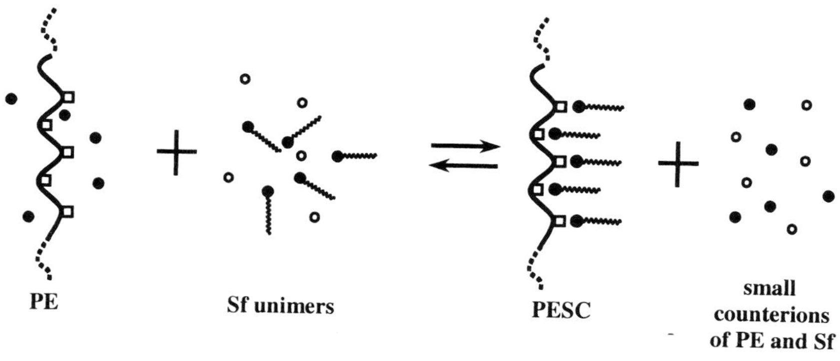

PE   **Sf unimers**   **PESC**   **small counterions of PE and Sf**

Scheme 1a

© 2004 WILEY-VCH Verlag GmbH & KGaA, Weinheim

Scheme 1b actually illustrates that in such case at $C_{Sf} <$ c.m.c. Sf ions have no reason to form a micellar phase replacing similarly charged polyions from the original IPEC, i.e. to transform IPEC into PESC.

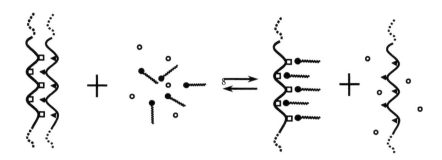

Scheme 1b

Indeed, if it were the case simple counter ions of Sf would become immobilised to neutralize the released polyions. In other words the number of statistically independent species would be diminished decreasing translational entropy, i.e. increasing free energy in the system, while hydrophobic interaction at $C_{Sf} <$ c.m.c. is still not sufficient to drive the process from left to right. In fact experimental data of Figure 1 show that #IPECs does not sorb Sf below c.m.c.

However, it must not be ruled out that in some special cases PESCs can be formed even at C_{Sf} somewhat below c.m.c. Hypothetically it may happen e.g. if one of IPEC constituents is of a low linear charge density. Then the original IPEC should contain a number of loops formed by low charge density polyions as shown in Scheme 1c. In such case reaction **1** may be driven from left to right by conformational entropy gain due to untangling the loops when low charge density polyion is released.

© 2004 WILEY-VCH Verlag GmbH & KGaA, Weinheim

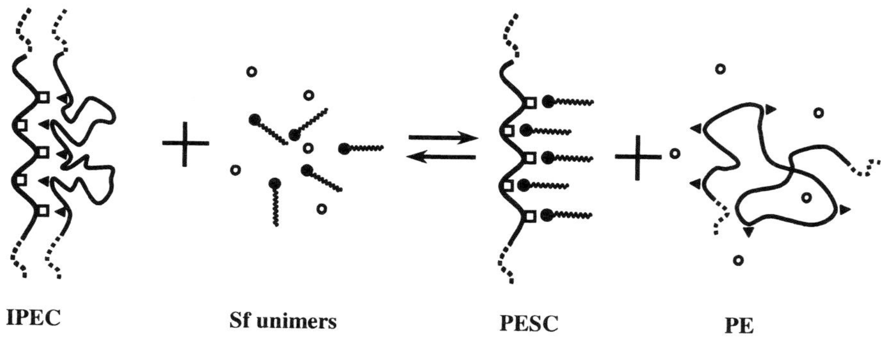

IPEC　　　　　**Sf unimers**　　　　　**PESC**　　　　　**PE**

Scheme 1c

In the ternary systems on study the situation when *I*PEs and Sf ions may actually compete for binding to #PE on equal terms appears at $C_{Sf} \geq$ c.m.c., i.e. when free Sf micelles are already formed as illustrated by Scheme 1d.

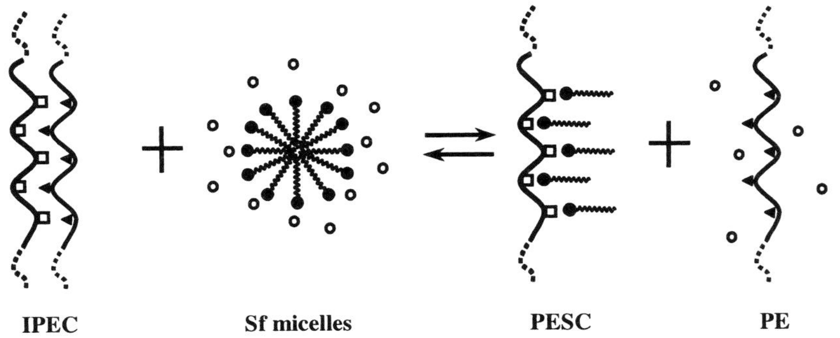

IPEC　　　　　**Sf micelles**　　　　　**PESC**　　　　　**PE**

Scheme 1d

In thermodynamic sense this situation resembles the earlier described and considered cases when two similarly charged *I*PEs[22, 23] or *I*PE and the vesicle[24] compete for binding to the oppositely charged *I*PE. In this case if the competitive reaction goes from left to right the counter ions of Sf micelles transfer to the *I*PE and vice versa. Apparently, the direction of the

© 2004 WILEY-VCH Verlag GmbH & KGaA, Weinheim

process therewith is determined by the relative affinity of Sf micellar phase and *I*PE to #PE and the counter ions to Sf micelles and *I*PE. The binding of counter ions by linear polyions predicted by Manning's theory is in a good agreement with the experiments.[25] Comparison of the literature data on the degree of binding of simple counter ions by *I*PE[25] and Sf[26] used in the present study shows that these values are quite congruent. Generally the ability of polyions to be either coupled if oppositely charged, or to bind counter ions increases with an increase in their linear charge density. Indeed, we observed that #IPECs containing PEI as *I*PE don't sorb different Sfs (systems 1-2, Table 2) even at Sf concentrations exceeding c.m.c. It can be referred to high charge density of short-branched PEI chains. On the contrary, PDADMAC polyions of a lower linear charge density in comparison with the fully charged vinyl type polyelectrolytes are not able to replace Sf ions from the studied #PESC (system 8, Table 1). Also note that #IPEC (#PA-PEVP) sorbs CTMAB releasing PEVPB but do not sorb DDTMAB, a lower Sf homologue. In other words equilibrium in Reaction (1) is shifted from left to right in system 7, Table 1 and from right to left in system 3, Table 2. The above difference could be caused by higher affinity of Br⁻ counter ions to DDTMA micelles as compared with CTMA. However, it was experimentally shown that ability of small ionic Sf micelles to bind counter ions only increases with an increase of Sf homologue number.[26] So the most probable reason is relatively enhanced stabilization of the structure of CTMA phase if reinforced by #PA as compared to DDTMA. The contributions of free *I*PE and Sf micelles in solution free energy should be also taken into account on quantitative treatment.

The kinetics of the above competing reactions in ternary #PE-*I*PE-Sf is of a primary interest. Figure 2a represents a typical kinetic curve of Sf uptake by #IPEC.

© 2004 WILEY-VCH Verlag GmbH & KGaA, Weinheim

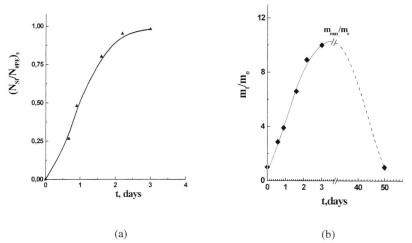

(a)   (b)

Figure 2. Kinetics of SDBS sorption (a) by the #IPEC (#PDMEAEM-PA) from an aqueous solution and kinetics of change in mass (m_t/m_o) (b) of the swollen stoichiometric #IPEC (#PDMEAEM-PA) sample; mass of the initial #IPEC sample 14 mg. [SDBS] = 0.005 mol/l, pH 8, T = 18 °C.

It is seen that at the given experimental conditions the sorption proceeds monotonously and $N_{Sf}/N_{\#PE}$ approaches unity and levels off for about 2.5 hours. It means that such time is sufficient for transformation of practically all salt bonds between #PE and *l*PE into the new salt bonds between #PE and Sf ions i.e. formation of stoichiometric #PESC. In other words charged groups of linear chains originally involved in cooperative electrostatic interaction with #PE, now are neutralized by simple counter ions of Sf. It is remarkable that in contrast to monotonous kinetics of Sf uptake the time dependence of mass (degree of swelling) of the reacting sample passes through a maximum. It is seen from Figure 2b that the originally compact #IPEC sample swells progressively when immersed into the Sf salt-free aqueous solution increasing its current mass (m_t) and reaching a maximum mass (m_{max}) one order of magnitude exceeding the mass (m_o) of the initial #IPEC sample. Then the mass gradually decreases and approaches of approximately original value for nearly 50 days.

The explanation is given taking into account that if reaction **1** proceeds from left to right, the influx of Sf unimers trough the surface of the reacting sample is much higher than efflux of

© 2004 WILEY-VCH Verlag GmbH & KGaA, Weinheim

*I*PE owing to the significant differences in their internal diffusion coefficients. Therefore by the time formation of #IPES is already completed most of the *I*PE chains still remains in the interior of the piece of the transformed hydrogel. In fact turbidimetric titration of environmental solution with PEVB that is quite sensitive method for detection of dissolved negatively charged polyions [17] showed that a measurable amount of *I*PE was released only in 3 days. However, now the chains temporally trapped in the network of the sample are not bound anymore to ionic groups of #PE, but neutralized by simple counter ions originally belonged to the Sf. These counter ions produce the essential part of additional osmotic pressure responsible for extra swelling of the hydrogel composite. All other systems listed in Table 1 behave in the same fashion, although differ in m_{max}/m_o values (Table 3).

Table 3. (m_{max}/m_o) values of the intermediate products of the interactions between #IPEC and Sf.

№	SYSTEM	C_{Sf}, mol/l	m_{max}/m_o
1	#IPEC (PA – PDADMA) + CPyC	0.006-0.002	40-50
2	#IPEC (PDMEAEM – PA) + SDBS	0.005-0.003	10-20
3	#IPEC (PDMEAEM – PA) + SDS	0.02-0.01	10-20
4	#IPEC (PA – PEVP) + CTMAB	0.10-0.01	5-10

The absolute contribution of *I*PE counter ions in the osmotic pressure depends on their osmotic coefficients, all factors being equal. The osmotic coefficient in its turn is determined by the polyion linear charge density and the ability of a polyion to bind counter ions specifically. Therefore the hydrogel composite containing PDADMA polycation, characterized by a relatively low linear charge density has a highest m_{max}/m_o value. At the same time the lowest m_{max}/m_o value is observed for the hydrogel composite containing PEVP, a polycation of a higher charge density and in addition to that characterised by a certain specific affinity to Br$^-$ ions (Table 3). As the free *I*PE slowly releases from the sample which at m_{max} is nothing but #PESC filled with a swollen *I*PE, the degree of swelling decreases in parallel. Here the release is simply driven by free energy decrease upon mixing *I*PEs with environmental water solution. We believe that ternary hydrogels of this kind approached

© 2004 WILEY-VCH Verlag GmbH & KGaA, Weinheim

$m_t = m_{max}$ may originate a novel perspective family of polymeric constructs applicable for gradual delivery of various polyelectrolytes oppositely charged relative to the #PE, such as proteins or nucleic acids. Earlier it was shown that either proteins[4, 5] or DNA[27] form #IPECs when interact with oppositely charged #PE. Apparently the efflux of releasing macromolecules i.e. the rate of release can be controlled, e.g. by degree of #PE cross-linking.

The kinetic behaviour of the ternary systems listed in Table 2 is entirely different. The equilibrium in competitive Reaction (1) for these systems is shifted from right to left. In other words the preformed #IPECs do not sorb Sfs. However, the preformed corresponding #PESCs immersed in water solutions of /PEs are able to uptake the /PE polyions releasing similarly charged Sf ions in equivalent amount. As an example the kinetics of the release of CPy cations from the stoichiometric #PESC (#PA-CPy) immersed in the solution of PEI·HCl is shown in Figure 3.

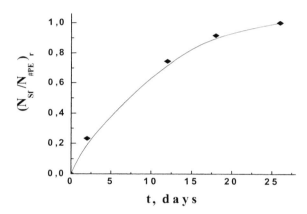

Figure 3. Kinetics of release of the CPy cations from a stoichiometric #PESC (#PA-CPy) sample into an aqueous solution of PEI·HCl; mass of the initial sample 20 mg, [PEI·HCl] = 0.1 mol/l, pH 4.2, T = 45 °C. N_{Sf} – is the mol amount of released Sf, $N_{#PE}$ – is the total base mol amount of #PE.

It is seen that here the rate of Sf ions release from the reacting sample is much less than their uptake in the reverse Reaction (1) (Figure 2a). No additional swelling similar to that shown in Figure 2b is observed: the reacting sample remains compact through out the whole process of Sf release. The above difference is not surprising and actually means that Sf ions are permitted

© 2004 WILEY-VCH Verlag GmbH & KGaA, Weinheim

to release from the reacting sample into surroundings only in response to polyions uptake. So the bottleneck of the process is slow internal diffusion of *I*PE. In so far as the internal diffusion coefficient for Sf unimers is much higher than polyions the release and uptake are synchronized, i.e. the reacting sample does not accumulate Sf at intermediate degrees of conversion in Reaction (1).

#IPECs either synthesized upon direct interaction of #PE hydrogels with water solution of oppositely charged non-stereoregular *I*PE or formed as a result of competitive Reaction (1) exhibit neither wide-angle nor small-angle X-ray ordering. Contrastingly a pronounced ordering is revealed in #PESC samples investigated by small angle X-ray scattering (SAXS). Figure 4 represents a typical SAXS diffractogramm for #PESC prepared via completed Reaction (1) between #IPEC (#PA-PEVP) and CTMAB, which is quite similar to that presented in our earlier paper [7] for the same #PESC obtained directly from #PANa and CTMAB.

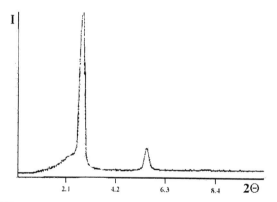

Figure 4. SAXS diffractogramm of the dry #PESC (#PA-CTMA) formed as a result of the complete reaction between #IPEC (#PA-PEVP) and CTMAB. T = 20 °C.

Referring to [7] the two observed peaks should be attributed the first and second order reflections from CTMAB lamellas. Interplanar spacings measured for the dry #PESCs samples prepared either by direct synthesis or via Reaction (1) are shown in Table 4.

© 2004 WILEY-VCH Verlag GmbH & KGaA, Weinheim

Table 4. Interplanar spacings (Å) in dry products of complete reactions between #IPEC and a Sf estimated from SAXS data.

SAMPLES	d_1	d_2
#PESC (#PA-CTMA)*	32.8	16.2
#PESC (#PA-CTMA)	31.6	16.0
#PESC (#PDMEAEM-DBS)	32.4	16.2
#PESC (#PDMAEM H-DS)*	36.8	-
#PESC (#PDMEAEM-DS)	35.6	17.8

*#PESC prepared via uptake of the Sf by oppositely charged #PE.

It is seen that the corresponding complexes are characterized by very close spacing values independently on the way they have been prepared. In our previous studies it was found that Sf ions incorporated in directly prepared #PESCs had a lamellar structure.[7] So it is worth to assume that lamellar structure is also typical for dry #PESCs obtained via completed Reaction (1). Having in mind the above considerations it is also worth to assume that swollen intermediate products of competitive Reaction (1) in particular those with a maximum degree of swelling are micro heterogeneous systems consisting of lamellar fragments electrostatically bound to the #PE network and distributed in the continuous phase composed of trapped water swollen /PE as it is shown in Scheme 2.

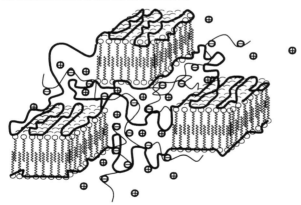

Scheme 2

© 2004 WILEY-VCH Verlag GmbH & KGaA, Weinheim

Conclusion

The above discussed ternary systems composed of #PE and the mixture of *I*PE and Sf both oppositely charged relative to #PE may be considered as intelligent multicomponent polymeric constructs. They are able to exchange the soluble species such as *I*PE and Sf with surroundings. The polyelectrolyte network is able to release linear polyions into surroundings in respond to uptake of Sf ions. Importantly the base mol amount of *I*PE released equals to the mol amount of sorbed Sf ions. The release rate is determined by characteristics of *I*PE mainly its molecular mass and linear charge density. Sf controlled release of polyelectrolytes from #IPECs may be of a particular interest with respect to proteins, polynucleotides and polysaccharides.

[1] V. A. Kabanov, A. B. Zezin, V. B. Rogacheva, V. A. Prevish, *Makromol. Chem.* **1989**, *190*, 2211.
[2] A. B. Zezin, V. B. Rogacheva, V. A. Kabanov, *J. of Intelligent Material Systems and Structures* **1994**, *5*, 1, 144.
[3] A. B. Zezin, V. B. Rogacheva, V. A. Kabanov, *Macromol. Symp.* **1997**, *126*, 123.
[4] V. B. Karabanova, V. B. Rogacheva, A. B. Zezin, V. A. Kabanov, *Polym. Sci. A* **1995**, *37*, 11, 1138.
[5] V. B. Skobeleva, V. B. Rogacheva, A. B. Zezin, V. A. Kabanov, *Dokl. Acad. Nauk (Russia)* **1996**, *347*, 2, 207.
[6] Yu. V. Khandurina, V. B. Rogacheva, A. B. Zezin, V. A. Kabanov, *Polym. Sci. B* **1994**, *36*, 2, 184.
[7] Yu. V. Khandurina, V. L. Alexeev, G. A. Evmenenko, A. T. Dembo, V. B. Rogacheva, A. B. Zezin, *J. de Phys. II France*, **1995**, *5*, 337.
[8] V.R. Ryabina, S. G. Starodoubtzev, A. R. Khokhlov, *Polym. Sci.* **1990**, *32*, 5, 903.
[9] A.R. Khokhlov, S. G. Starodoubtzev, V. V. Vasilevskaya, *Adv. Polymer Sci.* **1993**, *109*, 123.
[10] V. A. Kabanov, A. B. Zezin, V. B. Rogacheva, Yu. V. Khandurina, O. A. Novoskoltseva, *Macromol. Symp.* **1997**, *126*, 79.
[11] P. Hansson, *Langmuir* **1998**, *14*, 2269.
[12] O. A. Novoskoltseva, T. V. Krupenina, S. N. Sul`yanov, N. N. Bel`chenko, V. B. Rogacheva, A. B. Zezin, V. A. Kabanov, *Polym. Sci. A* **1997**, *39*, 7, 760.
[13] V. A. Kabanov, A. B. Zezin, V. B. Rogacheva, O. A. Novoskoltseva, T. V. Krupenina, *Dokl. Phys. Chem. (Russia)* **1998**, *358*, 4-6, 55.
[14] V. A. Kabanov, A. B. Zezin, V. B. Rogacheva, O. A. Novoskoltseva, *Dokl. Chemistry (Russia)* **2000**, *373*, 1-3, 121.
[15] K. N. Bakeev, V. A. Izumrudov, N. A. Sadovski, A. B. Zezin, M. G. Kuzmin, V. A. Kabanov, *Dokl. Acad. Nauk (Russia)* **1987**, *294*, 107.
[16] A. B. Zezin, N. M. Kabanov, A. I. Kokorin, *Vysokomol. Soedin. (Russia)* **1977**, A, *19*, 7, 118.
]17] V. A. Kabanov, A. B. Zezin, M. I. Mustafaev, V. A. Kasaikin, in *"Polymeric Amines and Ammonium Salts"*, E. J. Goethals, Ed., Pergamon Press, Oxford, New York 1980, p 173.
[18] Yu. V. Zanevskii, D. E. Donets, A. B. Ivanov, S. A. Movchan, A. I. Ostrovnoi, S. P. Chernenko, D. M. Kheiker, M. E. Andrianova, A. N. Popov, S. N. Sul`yanov, *Crystallography (Russia)* **1993**, *38*, 2, 252.
[19] K. Shinoda, T. Nakagawa, B. Tamamushi, T. Isemura, *"Colloidal Surfactants"*, Academic Press, New York 1963.

© 2004 WILEY-VCH Verlag GmbH & KGaA, Weinheim

[20] A. A. Abramzon, *"Poverkhnostno-aktivnye veshchestva"*, Khimiya, Leningrad (Russia) 1979.

[21] E. D. Goddard, *Interactions of Surfactants with Polymers and Proteins*, Boca Raton, CRC, **1993**.

[22] V. A. Izumrudov, T. K. Bronich, O. S. Saburova, A. B. Zezin, V. A. Kabanov, *Makromol. Chem., Rapid. Commun.*, **1988**, *9*, 7.

[23] A. B. Zezin, V. A. Izumrudov, V. A. Kabanov, *Frontiers of Macromol. Sci.*, **1989**, 219.

[24] A. A. Yaroslavov, E. G. Yaroslavova, A. A. Rakhnyanskaya, F. M. Menger, V. A. Kabanov, *Colloids and Surfactants*, **1999**, *16*, 29.

[25] Manning G. S., *J. Chem. Phys.*, **1969**, *51*, 924.

[26] R. Kakehashi, Y. Kanakudo, A. Yamamoto, H. Maeda, *Langmuir*, **1999**, *15*, 4194.

[27] V. G. Sergeyev, O. A. Novoskoltseva, O. A. Pyshkina, A. A. Zinchenko, V. B. Rogacheva, A. B. Zezin, K. Yoshikawa, V. A. Kabanov, *J. Am. Chem. Soc.*, **2002**, *124*, 11324.

© 2004 WILEY-VCH Verlag GmbH & KGaA, Weinheim

Macromol. Symp. **2004**, *211*, 175-189

Adsorption of Linear and Star-Shaped Polyelectrolytes to Monolayers of Charged Amphiphiles

M. Schnitter, H. Menzel*

Institut für Technische Chemie, Technische Universität Braunschweig,
Hans-Sommer-Straße 10, 38106 Braunschweig, Germany
E-mail: h.menzel@tu-braunschweig.de

Summary: Adsorption of polyelectrolytes has been studied employing monolayers of ionic amphiphiles at the air water interface as model surfaces. The adsorption of polyelectrolytes from a solution brought into contact with the amphiphile monolayer results in changes of the monolayer structure and properties. Monitoring these changes can be done by recording the changes in surface pressure. The kinetics of the adsorption depends strongly on the nature of the polyelectrolyte. Depending on the structure of the polyelectrolyte a purely diffusion controlled adsorption or a sequence of diffusion controlled adsorption and ordering processes have been identified to determine the kinetics. The influence of the molecular architecture on the poly-electrolyte adsorption has been further studied employing linear and star shaped poly(acrylic acid) and poly(N-propyl-4-vinyl pyridinium bromides), respectively. An unexpected behavior with an induction period in the adsorption kinetics of both polymers has been found. Furthermore, the degree of branching has only very minor effects on the adsorption kinetics.

Keywords: adsorption kinetics; amphiphiles; monolayers; polyelectrolytes; star-shaped polyelectrolytes

Introduction

Polyelectrolyte adsorption onto solid substrates plays an important role in many applications such as modification of surfaces e.g. preparation of antistatic and protective coatings or making artificial organs biocompatible. But also many technical processes like wastewater treatment and paper production are closely connected with polyelectrolyte adsorption.[1] Although there is a big interest in the adsorption process, only a few methods are available for *in-situ* investigation of the polyelectrolyte adsorption onto charged surfaces like ellipsometry[2] or surface plasmon resonance spectroscopy.[3] Monolayers of ionic amphiphiles at the air water interface can be regarded as models for charged surfaces. The adsorption of polyelectrolytes from a solution brought into

© 2004 WILEY-VCH Verlag GmbH & KGaA, Weinheim

DOI: 10.1002/masy.200450712

contact with the amphiphile monolayer results in changes of the monolayer structure and properties. Monitoring these changes can be done by different methods and should allow to obtain information about the adsorption process.[4, 5]

Polyelectrolyte-amphiphile complexes are formed spontaneously when an amphiphile solution is spread on a subphase containing the polyelectrolyte.[6, 7] Complex formation may results in a stabilization of monolayers and LB films of ionic amphiphiles, or in a change in the their structure[6, 7] or morphology.[8] Studying the monolayer behavior of the complexes allows to obtain information about the interaction between polyelectrolyte and amphiphile and about the new type of structure that has been formed. In addition the adsorption process can be monitored if preformed monolayers are brought into contact with a polyelectrolyte containing subphase.[9, 10]

There are several methods to prepare a monolayer of an ionic amphiphile on pure water and to bring it subsequently into contact with a polyelectrolyte solution. The polyelectrolyte can be added to the water subphase after compression of the monolayer as a concentrated solution.[11] Drawbacks of this method are that the subphase has to be stirred in order to ensure a homogeneous distribution of the polyelectrolyte and that the polyelectrolyte has to be added as highly concentrated solution. These problems can be avoided if the subphase is continuously exchanged.[12, 13] However, this methods requires a special trough and a fine control of supply and discharge of the subphase, in order to keep the level in the trough constant. Another drawback is that the flow has to be rather low in order to keep the mechanical stress on the monolayer as low as possible. However, at low flow rates there is a laminar flow profile and a concentration gradient perpendicular to the flow direction occurs. Therefore, long exchange times are necessary.[13]

Fromherz has reported a method in which the monolayer is prepared on one compartment of a multi-compartment trough which contains pure water and is subsequently transferred to another compartment which contains a polyelectrolyte or enzyme solution [14, 15, 16, 17] (s. Figure 1). The water level in the compartments is slightly higher than the walls, but due to the meniscus formed at the hydrophobic walls there is no contact between the solutions. During the transfer process, the contact area between the monolayer and the polyelectrolyte solution increases linearly. A minor mixing of the subphases during the transfer cannot be avoided and the transfer speed is limited by the stability of the monolayer.

© 2004 WILEY-VCH Verlag GmbH & KGaA, Weinheim

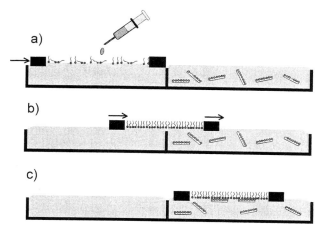

Figure 1. Fromherz method for bringing a preformed monolayer into contact with a subphase containing a polyelectrolyte: a) spreading the amphiphile solution on a compartment with pure water and subsequent compression of the monolayer, b) transfer of the compressed monolayer onto a compartment that contains a polyelectrolyte solution, and c) monitoring the adsorption process

Measuring the surface pressure is a well established method and can be performed easily even on a two compartment trough. So, recording the changes in the surface pressure after transferring the monolayer to the polyelectrolyte containing subphase is a straightforward way to monitor the adsorption. The expected change in surface pressure can be estimated from the comparison of the isotherms recorded on pure water and on a polyelectrolyte subphase (s. Figure 2). Experimental details for this method are given in Ref. [5, 18, 19].

Materials

The amphiphiles forming the model monolayer have to be chosen very carefully. The monolayer has to have sufficient stability to be transferred on the Fromherz trough. Furthermore, the changes in the structure of the monolayer upon adsorption of the polyelectrolyte have to be in such a way that they can be monitored.

Dioctadecyldimethylammonium bromide (DODA) is an amphiphile that is well suited to investigate the adsorption process for anionic polyelectrolytes[19, 20, 21] The isotherm of DODA (Figure 2) shows two regions, an expanded monolayer structure at high area per amphiphile,

© 2004 WILEY-VCH Verlag GmbH & KGaA, Weinheim

which turns into a more condensed structure at a surface pressure of approximately 20 mN/m.[22,23,24] The DODA monolayers are stable and suitable for the transfer according to the Fromherz method in both regions.[12,14] The area requirements for DODA on pure water are given by the coulomb interactions of the ionic head groups and the packing of the alkyl chains.[20]

DODA

DMPA

The isotherms for complexes of DODA and the anionic polyelectrolytes like poly(styrene sulfonate) (PSS),[21] carboxymethylcellulose sodium salt (CMC)[21] and polyacrylic acid (PAA)[18] (for the structure of the polymers s.Table 1) are significantly different from the isotherm of DODA on pure water. For complexes with PSS the area per DODA amphiphile is reduced to 0.60 nm^2.[21] The area per amphiphile in the complex is close to the value found in single crystals (0.56 nm^2 [25]). The lower area per amphiphile in the complex monolayer compared to the monolayer on pure water can be ascribed to the charge compensation, which significantly reduces the coulombic interaction of the amphiphiles head groups.

The area per amphiphile is larger for the complex of DODA with PAA complex than for the pure DODA monolayer. For PAA a incorporation of some non ionised carboxylic acid groups into the monolayer has been made responsible for the larger area per DODA molecule found for the complex.[18] It is well known, that PAA shows some surface activity.[26] Furthermore it has been shown by Ishimuro and Ueberreiter that only the protonated carboxylic acid groups are surface active.[27] It is therefore expected that the area per amphiphile in the complex depends on the pH of the subphase. This is in fact the case, as can be seen in Fig. 2. The lower the pH, that is the higher the degree of protonation, the higher is the area per amphiphile in the complex. Indeed at high pH, when all carboxylic acid groups are deprotonated the area per amphiphile is lower for the complex than for DODA on pure water. Thus for a not to high pH a significant change in the surface pressure can be expected when a DODA monolayer prepared on a pure water subphase is brought into contact with a polyelectrolyte solution as subphase.

© 2004 WILEY-VCH Verlag GmbH & KGaA, Weinheim

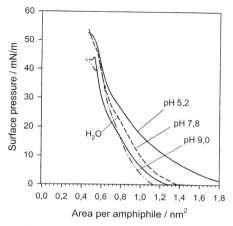

Figure 2.   Comparison of the isotherms for DODA spread on pure water, and solutions containing PAS at pH = 5.2 (12 mg/L PAA dissolved in ultrapure water), at pH = 7.8 (12 mg/L PAS dissolved in ultrapure water and 100 µl 1 M NaOH added to 210 ml subphase) at pH = 9.0 12 mg/L PAS dissolved in ultrapure water and 150 µl 1 M NaOH added to 210 ml subphase)

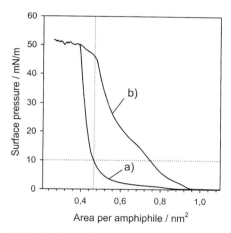

Figure 3. Comparison of the isotherms for DMPA spread on a) pure water, b) on an aqueous solution of poly(N-methyl-4-vinylpyridium bromide) q1PVP (c = 12.2 mg/L)

DMPA is a two chain anionic amphiphile which shows sufficient stability for the transfer in the Fromherz technique and the isotherms on pure water and on cationic polyelectrolytes like poly(N-methyl-4-vinylpyridium bromide) (q1PVP) are significantly different (Figure 3). Again the

© 2004 WILEY-VCH Verlag GmbH & KGaA, Weinheim

area per amphiphile is larger for the polyelectrolyte complex compared to the monolayer on pure water. This is believed to be due to the incorporation of hydrophobic parts of the polyelectrolyte into the monolayer. A large change in the surface pressure can be expected when a DMPA monolayer is brought into contact with a polyelectrolyte solution.

Table 1. Polyelectrolytes used for the investigations

polyelectrolyte	structure	molecular weight
sodium salt of carboxymethylcellulose, (CMC)	$R = -CH_2COONa$ $DS = 0.7$	$M_w \sim 90\ 000$ g/mol[1]
sodium salt of poly(styrol-p-sulfonate), (PSS)	SO_3Na	$M_w \sim 100\ 000$ g/mol[2]
sodium salt of poly(p-phenylene sulfonate) (PPPS)	SO_3Na	$P_n = 4, 36, 54$ see ref. [19, 20]
polyacrylic acid linear, 4-arm, 6-arm, 12-arm star (PAA)		linear 83 000g/mol [3,5] 4-arm 114 000 g/mol [3,5] 6-arm 111 000 g/mol [3,5] 12-arm 54 000 g/mol [3,5]
quaternized poly(vinylpyridine) linear, 4-arm, 6-arm, 12-arm (R = methyl (q1PVP), R = propyl (q3PVP))	$R = -CH_3,$ $R = -C_3H_7$	linear 89 000 g/mol [4,6] 4-arm 95 000 g/mol [4,6] 6-arm 91 000 g/mol [4,6] 12-arm 118 000 g/mol [4,6]

[1]Aldrich, [2]Acros, [3]M_w determined by SEC, [4]M_w determined by light scattering, [5,6] determined for the non-ionic prepolymers [5]poly(tert.-butylacrylate) and [6]poly(vinylpyridine)

© 2004 WILEY-VCH Verlag GmbH & KGaA, Weinheim

The conformation of a flexible polyelectrolyte in solution depends on several factors, among which the charge density and the salt concentration are most important. The repulsion of the charges along the polyelectrolyte chain results in a stretching and an expanded coil is formed. The polymer can be described by the worm-like chain model. However, when the charges are screened or compensated the polymer forms a more compact random coil.[1] Therefore besides poly(styrene sulfonate) and carboxymethylcellulose,[8, 19] which show these changes in their conformation we used polyelectrolytes which have a more fixed conformation.

Polymers with a poly(p-phenylene) backbone have been previously used as model for the completely extended chain situation. Poly(p-phenylene sulfonate) (PPPS) was employed as an anionic polyelectrolyte forming complexes with cationic DODA.[5, 19, 20, 21] Star-shaped polymers have a more globular shape than linear chains. Therefore, in this work we used star-shaped polymers with varying number of arms as approximation of the random coil situation of the polyelectrolytes.

© 2004 WILEY-VCH Verlag GmbH & KGaA, Weinheim

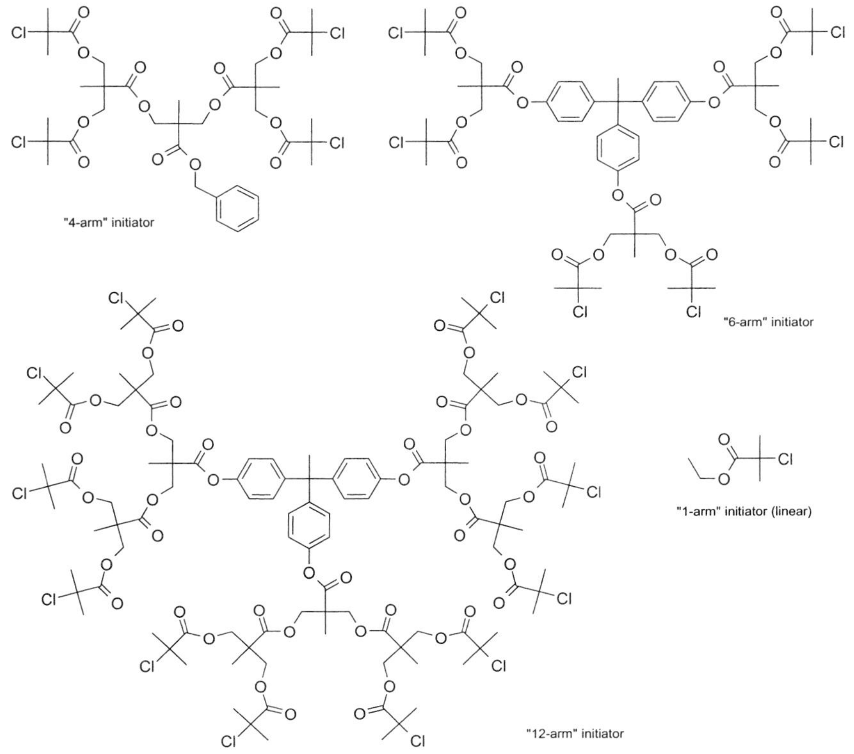

Figure 4. Dendritic initiators used for the synthesis of star shaped poly(vinylpyridine)s

We have synthesized polyacrylic acid[18] and poly(vinylpyridine) by ATRP employing dendritic initiators. The experimental details for the synthesis of the initiators and of star-shaped polyacrylic acid are given in ref. [18]. For the synthesis of the poly(vinylpyridine)s the same initiators were used but with chlorine end groups (s. Figure 4). Details of the synthesis are described in ref. [28].

© 2004 WILEY-VCH Verlag GmbH & KGaA, Weinheim

Surface Pressure/Area Isotherms for the Polyelectrolyte-Amphiphile-Complexes

The isotherms for DODA spread on subphases containing the different PAAs are shown in Figure 5. The isotherms are dissimilar for the different architectures of PAA. There is an increase in the area per amphiphile going from linear to 4-arm or 6-arm PAA. This trend is expected if an insertion of the hydrophobic core is assumed. However, the trend is not continued for the 12-arm PAA which has the largest hydrophobic core, but the area per amphiphile is only slightly higher than for the linear PAA. This unexpected behavior is can be correlated with the pH which is reached when the corresponding PAA is dissolved in pure water. The pH is the lowest for the 12-arm PAA (pH = 4.95) followed by the linear PAA (pH = 5.2), but higher for the 4-arm and 6-arm (both pH = 5.4). This slight differences in the pH of the solution might have a small but significant impact on the structure of the monolayer (Figure 2).

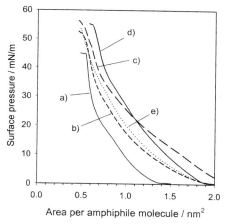

Figure 5. Surface pressure / area isotherms of DODA spread on a) pure water and solutions containing b) linear (c = 13.2 mg/L, pH = 5.2), c) 4-arm (c = 11.8 mg/L, pH = 5.4), d) 6-arm (c = 12.2 mg/L pH = 5.4), and e) 12-arm polyacrylic acid (c = 11.6 mg/L pH = 4.95)

© 2004 WILEY-VCH Verlag GmbH & KGaA, Weinheim

The cause for the different pH of the PAA solutions has not yet been identified unambiguously, but may be related to the differences in ionisation behavior of linear and star shaped molecules. However, theories would predict a reduced ionization for the higher branched weak polyelectrolyte, that is a higher pH for the 12-arm polymer.[29, 30]

Poly(N-propyl-4-vinylpyridinium bromide) (q3PVP) is a strong polyelectrolyte which degree of ionisation does not depend on the pH of the solution. The isotherms for linear and star-shaped q3PVP are shown in Figure 6. The same trend as with the star shaped PAA can be observed. The higher the degree of branching the larger is the area per amphiphile. Furthermore, the trend is continued for the 12-arm q3PVP. Thus in this case there is just the expected increase due to the larger hydrophobic core.

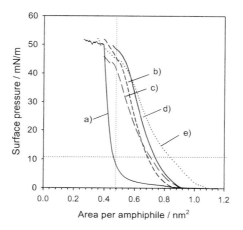

Figure 6. Surface pressure/area isotherms of DMPA spread on a) pure water and solutions containing b) linear (c = 12.8 mg/L), c) 4-arm (c = 12.2 mg/L), d) 6-arm (c = 11.8 mg/L), and e) 12-arm quarternized poly(vinylpyridine) (c = 12.1 mg/L)

Adsorption Experiments

The adsorption of the polyelectrolyte to an amphiphile monolayer brought into contact with the polyelectrolyte solution in Fromherz experiment can be monitored by recording the surface pressure as function of time (π-t-isotherms).[5, 18, 19, 31] It has been shown previously that there are significant differences between the π-t-isotherms for the different polyelectrolytes, while in the

© 2004 WILEY-VCH Verlag GmbH & KGaA, Weinheim

case of PSS and CMC, the surface pressure change seems to be a direct measure for the adsorption and the π-t-isotherms show a concentration dependence as expected for a diffusion controlled process,[19] this is not the case for PPPS. The π-t-isotherms show an inflection. Reducing the concentration of the polyelectrolyte in the subphase or increasing the molecular weight, the inflection becomes more pronounced and an induction period is found in which no change in the surface pressure takes place (Figure 7). In this case the change in the surface pressure is obviously no longer a direct measure of the adsorbed amount, but additional effects like the interactions of the polymer chains with each other, with the interface itself, and with the amphiphiles at the interface have to be taken into account.[19, 32, 33] Indeed it was possible to describe the dynamics of the adsorption process assuming a two-dimensional "crystallization" process.[19] Employing this approach the π-t-isotherms can be fitted nicely with an expression derived from the AVRAMI-equation, which has been developed to describe crystallization behavior of polymers.[34]

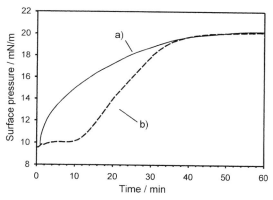

Figure 7. Comparison of the π-t-isotherms a) expected for an diffusion controlled adsorption (schematic) and b) those found for PPPS(P_n = 36, C = 20.0 mg/L (0,8·10^{-5} mol/L ionic sites)[19]

According to the results with PSS and CMC on one side and the rigid rod like polymers PPPS on the other side the adsorption process has to be divided in several steps: (1) the diffusion of the polymer chain to the monolayer, (2) the change in conformation and adsorption, (3) organization or "crystallization". For PPPS – that is the model for the extended chain situation – it has to be assumed that the ordering (step 3) is the rate determining step.[19]

© 2004 WILEY-VCH Verlag GmbH & KGaA, Weinheim

As mentioned already star-shaped polymers have a more globular shape than linear chains. It can be anticipated, that these more globular shaped polymers do not show ordering phenomena upon adsorption as observed for the rigid rod polymers. However, as shown in Figure 8 the π-t-isotherms for the different polyacrylic acids also show a kind of induction period, in which the surface pressure even is reduced compared to the original pressure.

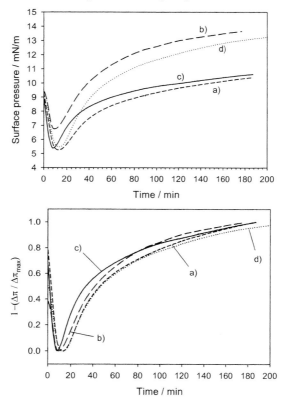

Figure 8. Change in surface pressure (π) upon transfer of a DODA monolayer onto a solution containing a) linear (c = 2.0 mg/L), b) 4-arm (c = 1.8 mg/L), c) 6-arm (c = 3.1 mg/L), and d) 12-arm polyacrylic acid (c = 1.5 mg/L) as function of time (top) and normalized change in surface normalized to the maximum change in surface pressure (bottom) (transfer of the monolayer at 10 mN/m)

© 2004 WILEY-VCH Verlag GmbH & KGaA, Weinheim

The extent of the surface pressure drop seems not to depend on the number of arms. It is somewhat less pronounced for the 6-arm polyacrylic acid, but for the linear, the 4-arm-, and the 12-arm-polymer it is almost the same. Furthermore, the π-t-isotherms show only a slight scattering when normalized to the maximum surface pressure change (s. Figure 8 bottom). Thus, there is an unexpected induction period for PAA, which does not significantly depend on the degree of branching.

The situation for the branched q3PVP is very similar. Again all polyelectrolytes show an unexpected behavior in the adsorption kinetics with an induction period in which the surface pressure drops. This is unexpected since the comparison of the π-A-isotherms suggest an increase in the surface pressure and no ordering phenomena are anticipated for q3PVP as it is the case for PPPS. Furthermore, the π-t-isotherms for the star-shaped polymers are very similar. Thus, there are no differences in the adsorption kinetics due to the degree of branching. However, for this class of polymers the linear polymer shows a different behavior (Figure 9), but the difference is merely quantitative than qualitative.

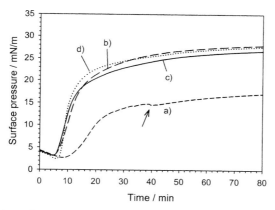

Figure 9. Change in surface pressure (π) upon transfer of a DMPA monolayer onto a solution containing a) linear (c = 1.9 mg/L), b) 4-arm (c = 1.8 mg/L), c) 6-arm (c = 1.9 mg/L), and d) 12-arm poly(N-propyl-4-vinylpyridium bromide) (c = 1.9 mg/L) as function of time (transfer of the monolayer at 5 mN/m), the arrow marks a partial collapse

The π-t-isotherm for the linear q3PVP has the same shape than those for the star-shaped q3PVP, but the final surface pressure is somewhat lower and the induction period slightly longer. The origin of the quantitative differences remain elusive. It can be speculated that a partial collapse of the monolayer upon adsorption and the concomitant increase in surface pressure limits the final surface pressure. The π-t-isotherm for the linear q3PVP show an indication for a partial collapse (discontinuity at approximately 45 min marked by an arrow in Figure 9).

The investigation of linear and star-shaped poly(acrylic acid) and poly(N-propyl-4-vinylpyridium bromide) indicate that the influence of the degree of branching on the kinetics of adsorption of the polyelectrolyte to a monolayer of charged amphiphile is not very significant. Furthermore, for the DODA/PAA as well as for the DMPA/qPVP a significant induction period is found in the π-t-isotherms, which is completely unexpected. The explanation for this behavior remains elusive at this moment, but it can be supposed that beside ordering effects also hydrophobic interaction may play a role in the adsorption process. The surface pressure change can be due to a incorporation (partly or completely) of the polyelectrolyte into the monolayer. This process is most likely the rate determining step after diffusion and adsorption.

Conclusions

We have shown that it is possible to use monolayers of ionic amphiphiles as model surfaces to study the polyelectrolyte adsorption. Previously rigid rod-like polymers were investigated as model for completely extended chain situation. Star-shaped polymers have a more globular shape than linear polymers and can be regarded as approximation of the random coil situation of the polyelectrolytes. In this study linear and star-shaped poly(acrylic acid) and poly(n-propyl-4-vinylpyridium bromide) with varying number of arms were used to study the influence of an increasing degree of branching on the adsorption kinetics. Previous results with rigid rod-like polyelectrolytes have indicated that the adsorption process is a multi-step process in which the adsorption itself is followed by an ordering process in the complex monolayer. Such an ordering process resulting in a liquid crystalline phase can be ruled out for a globular system. However, the experimental results with the linear and star-shaped polyelectrolytes clearly show that the also highly branched 12-arm polyelectrolytes can show an induction period in the adsorption kinetics measured as π-t-isotherm. The influence of the degree of branching on the adsorption kinetics seems to be small for both polyelectrolytes and merely quantitative.

© 2004 WILEY-VCH Verlag GmbH & KGaA, Weinheim

[1]   H. Dautzenberg, W. Jaeger, J. Kötz, B. Philipp, Ch. Seidel, Stscherbina D. *"Polyelectrolytes - Formation, Characterization and Application"*, Carl Hanser Verlag, München **1994** .
[2]   H. Walter, C. Harrats, P. Müller-Buschmann, R. Jérôme, M. Stamm, *Langmuir* **1999**, *15*, 1260.
[3]   N. A. Kotov, I. Dékány, J. H. Fendler, *Adv. Mater.* **1996**, *8*, 637.
[4]   N. Kimizuka, T. Kunitake, *Coll. Surf.* **1989**, *38*, 79.
[5]   M. Schnitter, J. Engelking, H. Menzel, *Coll. Surf. A* **2002**, *198- 200*, 187.
[6]   C. Erdelen, A. Laschewesky, H. Ringsdorf, J. Schneider, A. Schuster, *Thin Solid Films* **1989**, *180*, 153.
[7]   M. Shimomura, T. Kunitake, *Thin Solid Films* **1985**, *132*, 243.
[8]   J. Engelking, H. Menzel, *Thin Solid Films* **1998**, *327-*, 90.
[9]   K. Miyano, K. Asano, M. Shimomura, *Langmuir* **1991**, *7*, 444.
[10] K. Asano, K. Miyano, H. Uki, M. Shimomura, Y. Ohta, *Langmuir* **1993**, *9*, 3587.
[11] M. Kawaguchi, M. Yamamoto, T. Kato, *Langmuir* **1998**, *14*, 2582.
[12] F. Eßler, *Dissertation, Universität Mainz* , 1998.
[13] K. Morawetz, J. Reiche, L. Brehmer, W. Jaeger, *Presented at the 7th European Conference on Organized Films - ECOF7, 13.9. - 18.9.98, Potsdam, Germany* 1998.
[14] P. Fromherz, *Biochem. Biophys. Acta* **1971**, *225*, 382.
[15] P. Fromherz, *Rev. Sci. Instrum.* **1975**, *46*, 1380.
[16] Z. Kozarac, A. Dhathathreyan, D. Möbius, *Eur. Biophys. J.* **1987**, *15*, 193.
[17] S. Sundaram, J.K. Ferri, D. Vollhardt, K.J. Stebe, *Langmuir* **1998**, *14*, 1208.
[18] M. Schnitter, J. Engelking, A. Heise, R. D. Miller, H. Menzel, *Macromol. Chem. Phys.* **2000**, *201*, 1504.
[19] J. Engelking, H. Menzel, *Eur. Phys. J.* **2001**, *5*, 87.
[20] J. Engelking, D. Ulbrich, H. Menzel, *Macromolecules* **2000**, *33*, 9026.
[21] J. Engelking, H. Menzel, D. Ulbrich, W. H. Meyer, K. Schenk-Meuser, H. Duschner, *Mater. Sci. Eng. C* **1999**, *8-9*, 29.
[22] A. B. Sieval, R. Linke, G. Heij, G. Meijer, H. Zuilhof, E. J. R. Sudhölter, *Langmuir* **2001**, *17*, 7554.
[23] D. M. Taylor, Y. Dong, C. C. Jones, *Thin Solid Films* **1996**, *284-*, 130.
[24] Z. Kozarac, R. C. Ahuja, D. Möbius, *Langmuir* **1995**, *11*, 568.
[25] T. Kunitake, *Angew. Chem. Int. Ed. Engl.* **1992**, *31*, 709.
[26] R. V. Talroze, T. L. Lebedeva, G. A. Shandryuk, N. A. Plate, N. D. Stepina, L. G. Yamusova, L. A. Feigin, *Thin Solid Films* **1998**, *325*, 232.
[27] Y. Ishimuro, K. Ueberreiter, *Colloid Polym. Sci.* **1980**, *258*, 1052.
[28] M. Schnitter, *Dissertation, TU Braunschweig* 2003.
[29] R. Israels, F.A.M. Leermakers, G.J. Fleer, *Macromolecules* **1994**, *27*, 3087.
[30] E.B. Zhulina, T.M., Borisov, O.V. Birshtein, *Macromolecules* **1995**, *28*, 1481.
[31] J. Engelking, M. Wittemann, M. Rehahn, H. Menzel, *Langmuir* **2000**, *16*, 3407.
[32] R. Vilanove, F. Rondelez, *Phys. Rev. Lett.* **1980**, *45*, 1502.
[33] R. Vilanove, D. Poupinet, F. Rondelez, *Macromolecules* **1988**, *21*, 2880.
[34] L. Mandelkern *"Crystallization of Polymers"*, McGraw-Hill Inc., New York 1964.

© 2004 WILEY-VCH Verlag GmbH & KGaA, Weinheim

Macromol. Symp. **2004**, *211*, 191-200

Polyelectrolytes on Block Copolymer Surfaces

Martin Brehmer,* Lutz Funk, Lars Conrad, Dirk Allard, Patrick Theato

Johannes Gutenberg University, Institute of Organic Chemistry, D-55099 Mainz, Germany
E-mail: mbrehmer@mail.uni-mainz.de

Summary: Soft lithography and properties of amphiphilic block copolymers are combined in a new technique for the generation of patterned substrates, which can be used in different ways as templates for further processing. In these processing steps the deposition of polyelectrolytes, metals and grafting from polymerizations are used for the construction of different structures.

Keywords: block copolymers; pattern; polyelectrolyte; soft lithography; structure

Introduction

In recent years a major interest in science focuses on size reduction and the build up of small and versatile structures in the nanometer range. A newer technique besides optical lithography emerged from the need for methods, which can be used in standard laboratories. Soft lithography[1] fulfills the requirements for such a technique because it is cheap and relatively easy to handle. We are combining the properties of block copolymers and soft lithography for the preparation of small structures.

The surface of block copolymers is made up by the phase with the lowest interfacial energy.[2] If the surrounding medium is changed this behavior can lead to a surface reconstruction.[3] The driving force for this reconstruction of the surface is the gain in surface energy due to the different interfacial energies of the polymer blocks. We want to use this effect in order to create hydrophilic hydrophobic patterns on films of amphiphilic block copolymers. For this purpose we combine the surface reconstruction of block copolymers with soft lithography (Figure 1).

© 2004 WILEY-VCH Verlag GmbH & KGaA, Weinheim

DOI: 10.1002/masy.200450713

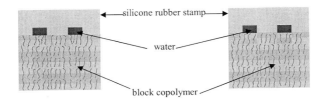

Figure 1. Scheme of surface patterning while (left) and after reorientation (right)

In order to obtain polymers with high mobility, poly(4-octylstyrene) with a Tg of –40 °C was chosen as hydrophobic block. The hydrophilic block will be prepared from 4-acetoxystyrene. The polymers were prepared by nitroxide controlled radical polymerization.[4] In the beginning a poly(4-octylstyrene) homopolymer is prepared. This can be used as a macroinitiator for the polymerization of 4-acetoxystyrene. The block copolymer can be deprotected by hydrazinolysis to give the amphiphilic poly(4-octylstyrene)-block-poly(4-hydroxystyrene) (1,2). As a second system, which can be switched at elevated temperatures, poly(styrene)-block-poly(acetic acid 2-[2-(4-vinyl-phenyl)-ethoxy]-ethyl ester) (3) was prepared.

Table 1: Properties of amphiphilic block copolymers

Polymer	M_w	PDI	ratio (hydrophilic/hydrophobic)	T_G hydrophobic phase °C	T_G hydrophilic phase °C
Poly(4-octylstyrene)-block-(4-hydroxystyrene)	53 000 63 000	1,44 1,48	10/1 (1) 1/1 (2)	-33	135
Poly(acetic acid 2-[2-(4-vinyl-phenoxy)-ethoxy]-ethyl ester)-block-(styrene)	63 000	1,36	1/2 (3)	101	-0,5
Poly(4-octylstyrene)-block-(2-[2-(4-vinyl-phenoxy)-ethoxy]-ethanol)	84 000	1,9	1/1,3 (4)	-24	14

The surface reconstruction of the polymer can be followed by changes of the contact angle. In the case of polymer (1) the molar ratio of the hydrophilic block is too high. The surface is already partly hydrophilic. The advancing contact angle drops only from 93° to 90° under water treatment. In the case of polymer (2) the surface is hydrophobic and the initial contact angle is 110°. During 24 h of water treatment, the contact angle drops down to 90°. The surface

© 2004 WILEY-VCH Verlag GmbH & KGaA, Weinheim

reconstruction is induced by the change of the surrounding medium. The water has a second crucial purpose in this process. Because of the high Tg of the hydrophilic polymer the reorientation should be hindered. The water acts as a softener due to the swelling of the hydrophilic block and so the Tg decreases below RT. If the water is removed and the sample is stored in air at room temperature, the contact angle does not switch back. This is due to the deswelling of the polymer, which causes an increase of Tg. The deswelling process is much faster than the reconstruction process and so the surface stays hydrophilic. If the sample is heated to 150 °C, which is above the glass transition temperature of poly(4-hydroxystyren), the value rises to 110° again.

A direct measurement of the change of the surface chemistry can be achieved by near edge x-ray absorption fine structure (NEXAFS) spectroscopy. In this x-ray experiment only the first few nanometers of the sample are probed. By excitation of the ^{1}s electrons into non binding orbitals, even the chemical surrounding of the excited atom can be determined. For the experiments polymer (3) were chosen because not only the oxygen content in the hydrophilic block is much higher. Also the carbonyl group has a strong and well distinguishable signal in NEXAFS measurements. The third reason for our choice is the difference of the concentration of the aromatic rings, which is quite large as well. The surface of freshly spin-coated polymer (3) is hydrophilic despite the fact that the hydrophobic block has the lower surface energy to air. To reach equilibrium the samples are tempered at 120 °C for 14 h and during this time the hydrophobic block appears at the surface. This can be seen in the high value for the aromatic C atoms. If the film is treated with water at 70 °C the surface reconstructs. An increase for the C-O and C=O bonds can be observed, whereas the signal for the aromatic C-atoms decreases. This can be explained by the exchange of the hydrophobic by the hydrophilic block. The same behavior can be found, if the total electron yield is measured (Figure 2).

© 2004 WILEY-VCH Verlag GmbH & KGaA, Weinheim

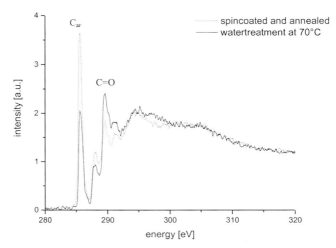

Figure 2. NEXAFS measurements of polymer **(3)** before and after reconstruction

If the sample is heated under nitrogen to 100 °C the values change back to the initial hydrophobic state.

The reorientation capability of amphiphilic block copolymers can be used to create hydrophilic hydrophobic structured interfaces. For the patterning process itself two different methods can be used. In the case of polymer **(4)** an exchange of the interface at room temperature can induce the reconstruction of the surface. This process can be started by using a structured hydrophilic poly (dimethylsiloxane) stamp (PDMS-stamp), which is brought into contact with the polymer surface. This stamp is created by pouring Sylgard 184 (Dow Corning) on top of a silicon master. After curing the mixture at 80 °C the stamp is hydrophilized in an oxygen plasma. Because of the low Tg of both polymer blocks the created surface pattern is not as stable as with polymer **(2)** mentioned above and will vanish in a much shorter time.

Polymer **(2)** can be structured with hydrophobic PDMS-stamps in the presence of water. The water creates the hydrophilic interface, which leads to the reconstruction in these areas. The PDMS covered parts remain hydrophobic and so a hydrophilic hydrophobic structured surface is obtained. The partially modified polymer film is still able to do the same reconstruction in further steps. In this way it is possible to obtain much more complicated structures. Another advantage of

© 2004 WILEY-VCH Verlag GmbH & KGaA, Weinheim

this method is that the structuring process works without any ink, which is in difference for a lot of other patterning techniques. Despite the advantage of the high stability of the pattern in contrast to the structuring process without water one major disadvantage exists. Because the water creates the hydrophilic interaction the structures on the stamp have to be connected so that water can flow into the created channels. This does not allow the fabrication of closed hydrophilic structures like spheres, squares or similar patterns.

Hydrophobic hydrophilic patterned surfaces, generated with this technique, feature OH groups in the hydrophilic parts. Beside the surface energies these OH groups can be used for a further chemical differentiations. This allows in addition to coulomb or van der Waals forces the build up of covalent bonds and the possibility to fix compounds permanently.

Because of the reconstruction process and swelling of the polymer the surface in the hydrophilic parts is not as smooth as it was initially. This height profile depends on the volume fraction of the used polymer blocks as it can be seen in (Figure 3). Polymers with higher content of the hydrophilic block show much larger height profiles. This can be understood because the swelling occurs in this block only and much more water can be absorbed.

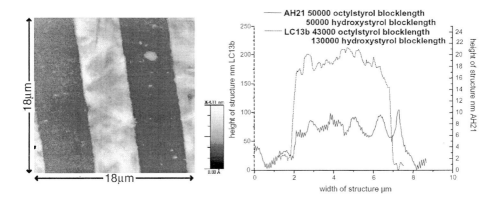

Figure 3. Height profile of 5μm structures in reliance on the polymer composition

© 2004 WILEY-VCH Verlag GmbH & KGaA, Weinheim

To deposit polyelectrolytes on a patterned block copolymer surface charges have to be generated. Because of the ability to deprotonate the phenolic groups of the block copolymer the reorientation was done in an alkaline solution. The coulomb forces inside the film lead to an increase of the height difference between the modified and unmodified polymer parts. This can be seen if an alkaline solution is used for the patterning process instead of water (Figure 4).

Figure 4. Height profile of a 5μm line prepared by reconstruction with alkaline solution **(2)**

To limit this effect it is possible to deprotonate the OH groups on the surface after reconstruction. For this purpose a structured substrate is dipped into a NaOH solution for a few seconds. To such a modified surface, which bears negative charges, polycations can easily be attached.[5] With this strategy it is possible to deposit cationic polyelectrolytes like polymer **(5)**[6] or **(6)** on a structured surface of polymer **(2)**. Another way to obtain various structures is to use a stamp to create a hydrophilic hydrophobic structured surface and then use a stamp as a mask for the following modification process. A similar pattern was created by Hammond et al.[7] They stamped the polyelectrolyte directly to the surface. Our microfluidic approach is more versatile because it does not depend on the interactions between the stamp and the polyelectrolyte. The deposition in squares of polyelectrolyte **(6)** is shown in Figure 5. A stamp with a line pattern was used to create a prestructured surface and after deprotonation with a 0.002M NaOH solution the stamp was put back onto the surface rotated at an angle of 90°. Now a 10^{-3} M solution of polyelectrolyte **(5)** in 1N HCl was brought in front of the lines and sucked into them by capillary forces. After 40 min the stamp was removed and washed three times with millipore water. It can be seen in (Figure 6) that the polyelectrolyte was deposited on the negatively charged parts.

© 2004 WILEY-VCH Verlag GmbH & KGaA, Weinheim

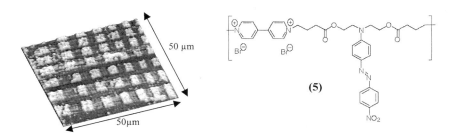

Figure 5. Deposition in squares of polyelectrolyte (5) on top of a 5µm line pattern (2) with a 90° turned stamp

Due to the azobenzene sidegroups polycation (5) might be used as an alignment layer for liquid crystals. As azo-dyes orient by irradiation with polarized light, it should be possible to control the director of the liquid crystal by the direction of the polarized light. This is known as the "command surface effect".[8]. The polymer pattern was covered with a commercial liquid crystalline mixture (ZLI 4431®, Merck). Figure 5 shows a polarized light micrograph of a sample of polymer (2) covered with squares of polycation (5) (Figure 6). A grid of planar aligned areas, which shows birefringence, surrounds dark squares with homeotropic alignment. This observation can be explained by the molecular structure of the liquid crystals. The liquid crystalline mixture mainly contains molecules with cyano end groups. This leads to a negatively charged head. If the molecules come into contact with a positively charged surface, the heads are attracted. The result is a homeotropic arrangement of the mesogens. In the case of a negatively charged surface the molecules seem to orient side on, because the aromatic rings carry a slightly positive charge. The same planar alignment is achieved with a hydrophobic surface. Due to these interactions with different surface areas, liquid crystals can be used to visualize surface patterns.

© 2004 WILEY-VCH Verlag GmbH & KGaA, Weinheim

Figure 6. Pattern from Figure 5 covered with an LC mixture. Squares with homeotropic alignment and lines with planar alignment can be distinguished. (scale is 50μm)

The deposition of a thiophene polyelectrolyte onto a line shape structured polymer film of polymer (2) is shown in (Figure 7). The most interesting property of this kind of polymer is the conductivity of the thiophenes. For the deposition of polyelectrolyte (6) the substrate with structured polymer (2) is dipped in a 0.1g/ml aqueous solution of sodium hydroxide for 5 seconds. After this the deposition itself is carried out with a 10^{-2} mol/l solution of the polyelectrolyte in water with 10% ethanol.

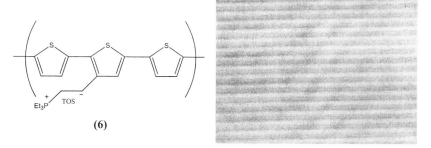

Figure 7. Deposition of thiophene polyelectrolyte on 5μm structured polymer (2)

© 2004 WILEY-VCH Verlag GmbH & KGaA, Weinheim

As mentioned before the OH groups on the surface of the polymer film are also accessible for chemical reactions in general. In addition to the mentioned deposition of polyelectrolytes, esterifications followed by grafting from polymerizations[9] or deposition of metals are done so far (Figure 8). The deposition of copper[10] for example is a multistep process in which first of all Sn^{2+} ions are coordinated to the oxygen atoms of the surface and in a second step Pd is deposited by reduction to the defined regions. The Pd atoms work as a catalyst for the copper deposition from a Cu^{2+} solution with formaldehyde. The produced copper pattern is very stable on top of the polymer film and it is not possible to remove it from the polymer film.

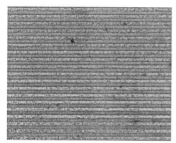 

Figure 8. Deposited copper left 5 μm lines of polymer **(2)** right: 10 μm triangles of polymer **(4)**

Conclusions

Amphiphilic block copolymers were prepared by nitroxide mediated polymerization of 4-octylstyrene and 4-acetoxystyrene. They can be patterned by bringing them in contact with a hydrophilic/hydrophobic interface. After deprotonation negatively charged surfaces are created. They can be decorated with polycations. Additional functionalities could be incorporated into the polymers. Polythiophenes allow the generation of patterns with conductive polymers. Liquid crystals could be used to visualize the surface pattern.

© 2004 WILEY-VCH Verlag GmbH & KGaA, Weinheim

[1] Y. Xia, G. M. Whithsides, *Angew.Chem.* **1998**, *110*, 586.
[2] G. Krausch, *Mater. Sci. Eng. R-Rep.* **1995**, *14*, 1.
[3] H. Mori, A. Hirao, S. Nakahama, *Macromolecules* **1994**, *27*, 4093.
[4] Mori et al., *Macromol. Chem. Phys.* **1994**, *195*, 3213.
[5] G. Decher, J. D. Hong, *Ber. Bunsenges. Phys. Chem.* **1991**, *95*, (11), 1430.
[6] A. Laschewsky, E. Wischerhoff, M. Kauranen, A. Persoons, *Macromolecules* **1997**, *30* (26), 8304.
[7] X. Jiang, P.T. Hammond, *Langmuir* **2000**, *16*, 8501.
[8] K. Ichimura, *Chem. Rev.* **2000**, *100*, 1847.
[9] M. Husemann, D. Mecerreyes, C. J. Hawker, J. L. Hedrick, R. Shah, N. L. Abbott, *Angew. Chem. Int. Ed.* **1999**, *38*, (5), 647.
[10] M. Charbonnier, M. Romand, E. Harry, M. Alami, *J. Appl. Electrochem.* **2001**, *31*, 57.

© 2004 WILEY-VCH Verlag GmbH & KGaA, Weinheim

Macromol. Symp. **2004**, *211*, 201-216   201

Comparison of Polyelectrolyte Multilayers Built Up with Polydiallyldimethylammonium Chloride and Poly(ethyleneimine) from Salt-Free Solutions by in-situ Surface Plasmon Resonance Measurements

Simona Schwarz,[1] *Jürgen Nagel,*[1] *Werner Jaeger*[2]

[1] Institut für Polymerforschung Dresden e.V., Hohe Str. 6, 01069 Dresden, Germany
Email: simsch@ipfdd.de
[2] Fraunhofer Institut für Angewandte Polymerforschung, Geiselbergstr. 69 14476 Golm, Germany

Summary: Polyelectrolytes offer a widespread potential for the defined modification of planar inorganic or polymer surfaces. Essential parameters for the regular adsorption of subsequent polymer layers by electrostatic interactions are the charge of polyelectrolyte and of the outermost surface region, the surface of the substrate, and the molar mass of the polyelectrolyte.
To study such effects in mono- and multilayers we used poly(diallyldimethylammonium chloride (PD) with a molar mass from 5000 to 400000 g/mol as a strong polycation and poly(ethyleneimine) (PEI) with 75000 g/mol as a weak polycation and poly(sodium styrenesulfonate) (PSS) from 70000 to 1Mio g/mol in the diluted and semi-diluted region.
The characterization of the layers was performed by streaming potential, in-situ SPR and UV-Vis spectroscopy. Thereby the layer built up at the solid/liquid-interface could be followed and quantified at the molecular level. SPR revealed that the thicknesses of the multilayer depends strongly on pK values of the polyelectrolyte (strong or weak) and the molar masses. We observed a linear growth if both polyelectrolytes are strong and an exponential growth if one polyelectrolyte is weak. The thickness increased with higher molar masses of the polyelectrolytes. The process was followed in-situ in short time steps.

Keywords: layer thickness; multilayers; polyelectrolyte adsorption; self-assembly; SPR; streaming potential

Introduction

Polyelectrolyte (PEL) multilayers are materials obtained by sequential adsorption of two or more polymers onto a substrate. The technique introduced by Decher [1] is based on the alternating dipping of the substrate into solutions of polyelectrolytes of opposite charge. The method is called "Electrostatic self-assembly" [2] because the primary forces for layer formation are the electrostatic interactions between polyelectrolytes and the substrate. However, secondary shorter range forces also play a role in determining such important parameters like film thickness, film morphology, surface properties and in

© 2004 WILEY-VCH Verlag GmbH & KGaA, Weinheim   DOI: 10.1002/masy.200450714

some cases, can even determine whether stable multilayers form at all.[3] Because of the versatility and simplicity of the technique, there has been a fast growth of publications in this area.[3, 4] The proposed applications for the so prepared novel materials range from sensors,[5] particularly biosensors,[6] light emitting diodes,[7] non linear optical devices[8] and perm selective gas membranes,[9] to controlled release micro-capsules[10] and bioactive surfaces for implants and tissue engineering.[11]

Most studies so far focused on the equilibrium properties and behavior of single component and mixtures of polyelectrolyte.[12-22] These types of studies are crucial in order to understand the thermodynamics involved, and to provide means for determining the amount of material adsorbed. As the adsorption process is mainly driven by electrostatic forces, the surface charge density and the net charge of the polyelectrolyte molecules play the dominant roles. Considering the system more in detail, the molecular structure of the surfaces,[22] as well as their composition and the nature of the exposed chemical functionalities come into play. Control of those details may offer opportunities for designing surfaces for special applications.

The need for characterization of polyelectrolyte layers has generated a broad band of modern analytical techniques (UV spectroscopy, SPR). Many analytical methods are useful for the characterization of PEL layers on smooth surfaces. The electrostatic interactions influence strongly the conformation of the adsorbed polyelectrolyte molecules. The interactions of the polyelectrolyte-coated substrates are determined by the layer structure, the distribution of the adsorbed segments and the electrostatic charge distribution. The characterization of the charge conditions of surfaces is possible with electrokinetic measurements.

Here we report the formation of some layer-by-layer thin films using PD as a strong cationic and PEI as a weak cationic polyelectrolyte and PSS as an anionic polyelectrolyte, in dependence on the molar mass on different substrates as followed by UV spectroscopy, streaming potential measurements, and in-situ surface plasmon resonance measurements (SPR).

© 2004 WILEY-VCH Verlag GmbH & KGaA, Weinheim

Experimental part

Materials

The poly(sodium styrenesulfonate) (PSS) and poly ethyleneimine (PEI) were purchased from Aldrich and Sigma, respectively, and used without further purification. The PEI is a highly branched product. The polydiallyldimethylammonium chloride (PD) is a linear product. The synthesis of PD is reported by Jaeger et.al.[23] Polymers with high (h) and low (l) molecular masses were used as follows: PSS(l) with Mw of 70 000 g/mol, PSS(h) with 1 Mio g/mol, PEI with 75 000 g/mol, PD(l) with 5 000 g/mol, and PD(h) with 400 000 g/mol. The polyelectrolyte concentrations for the adsorption studies were always 10^{-2} M related to the repeat unit of the polymer.

Water purified and deionized (reverse osmosis followed by ion exchange and filtration) by means of Milli-RO 5Plus and Milli-Q Plus systems (Millipore GmbH, Germany) was used as a solvent.

The quartz plates (Spektrosil B, Thermal Quarz-Schmelze GmbH, Germany) for investigations in the UV range had the dimensions of 76 mm x 26 mm x 1 mm. The plates were thoroughly cleaned by a mixture of sulfuric acid with potassium dichromate at 80 °C for about two hours in an ultrasonic bath prior to the deposition of such films for UV measurements.

For some experiments (surface plasmon resonance), high refractive index glass slides SF10 (Hellma Optic GmbH, Germany) covered by a thin evaporated gold layer with the thickness of 50 nm were used as the supports for the multilayers.

UV Spectroscopy

Multilayers were deposited from initial concentration c_0 of the cationic polyelectrolyte of 10^{-2} M related to the repeat unit of the polymer. The pH value of the solution was about 6. Adsorptions were carried out at room temperature in open glass beakers of 100 mL without stirring for 20 min. After every deposition layer, the substrates were rinsed three times (1 min each) with Millipore Milli-Q water. The substrates were not dried between the adsorption steps.

© 2004 WILEY-VCH Verlag GmbH & KGaA, Weinheim

Absorbance spectra of the dried films were measured by means of a Lambda 800 UV/VIS spectrometer (Perkin Elmer Ltd., USA). The multilayers were dried at room temperature in air atmosphere prior to the spectroscopic investigation.

Electrokinetic Measurements

The samples were prepared as those for UV spectroscopy. Electrokinetic measurements for the PEL layers were carried out by means of an Electrokinetic Analyzer device (A. Paar KG, Austria). The values of ζ-potential were calculated according to the formula:

$$\zeta = \frac{\eta}{\varepsilon_0 \varepsilon_r} \cdot \frac{\Delta U}{\Delta p} \cdot \kappa$$

where ΔU is streaming potential measured between two Ag/AgCl electrodes located at the opposite ends of the substrates, η, ε_r, κ are dynamic viscosity, relative dielectric permittivity, and conductivity of the flowing electrolyte solution (0.001 M KCl) respectively, ε_0 is dielectric permittivity of vacuum, and Δp is the pressure applied (150 mbar). The 0.1 M solutions of KOH and HCl were used to change pH of the flowing electrolyte solution in the range from 9 to 3.

Surface Plasmon Resonance

Surface plasmon resonance data for the layers were obtained by means of the equipment consisting of a He-Ne laser with $\lambda = 632.8$ nm (Uniphase, USA), a semi-cylinder made of SF10 glass, a liquid flow cell with the volume of 2.5 mL, and a E10V large area silicon photodiode detector with an integral preamplifier (Linus GmbH, Germany). The laser emitted polarized light (polarization ratio was 500:1) with power of 3 mW onto a semi-cylinder whose plane face was coupled, via index matching fluid, to the substrate examined (SF10 glass slide covered by the gold layer and then by the polyelectrolyte layers). The liquid flow cell was attached to the other side of the substrate and sealed with a rubber O-ring. The light was reflected onto the gold layer to excite surface plasmons. The intensity of the reflected light was measured by the photodiode. Both the semi-cylinder and the detector were mounted in an in-house $\theta/2\theta$ goniometer in such a way that the laser beam was incident on the detector at any angle of incidence. The goniometer and the photodiode were interfaced (MotionMaster 3000, Newport Corp.,

© 2004 WILEY-VCH Verlag GmbH & KGaA, Weinheim

USA) to a personal computer. An in-house 32-bit software package was used for goniometer control, data acquisition, curve modeling, and curve fitting. For scans over a certain range of incidence angles, a step width of 0.1° was used. The curves obtained were fitted according to Fresnel's equations for a four-layer model (glass/metal/dielectric/surrounding medium).

Multilayers were adsorbed in the measurements cell, and the measurements were done in-situ after adsorption onto the gold-covered glass plate. A peristaltic pump was used to circulate the solutions with a speed of 10 mL per minute. First, a SPR scan with the slide rinsed in water was performed. The pH of the PEI solution was pH = 9, all other polyelectrolyte solutions had pH of 6.5. Then a polyelectrolyte double layer was adsorbed according to the following procedure: circulation of the polycation solution, rinsing with pure water, circulation of the polyanion through the cell, and rinsing with pure water again. Then a SPR scan was performed. The procedure was repeated for each double layer.

All measurements were made at room temperature.

Results and Discussion

The adsorption of a polyelectrolyte molecule is mainly driven by electrostatic attraction to an oppositely charged surface. The adsorbed amount at the fist adsorption step must be so high that a charge overcompensation takes place which allows for the adsorption of unlike charged polyelectrolytes in the second step. It was often found[24] that the amount of adsorbed charges per adsorption step is approximately twice as much as the charge of the surface, half of which is attracted by the surface, the other half is repelled and distributed in the proximity of the surface. The density profile in this range depends on the solvent conditions. Repeating the adsorption steps would thus result in a constant adsorbed amount in each adsorption step. This result was found for a number of different polyelectrolytes, adsorbed under different conditions and measured with different approaches, ex-situ as well as in-situ.[25-30]

The first parameter to analyze is the charge density of the substrate used. The analytical techniques used to characterize the PEL layers required different substrates: quartz for UV spectroscopy and glass covered with a thin gold layer for SPR. The charge densities were characterized by the zeta potential which was measured over a certain range of pH values. The dependence of the zeta potential on pH for glass modified with a thin gold layer, and quartz glass are shown in Figure 1. The intersections with the lines at the working pH values (9 for PEI and 6.5 for the other) give the zeta potentials at which the adsorption occurred. All zeta potentials are negative at the working pH values. Consequently, multilayer built-up should start with a polycation as the first layer. The zeta potentials of quartz are close together at both pH values (approximately 40 mV and 45 mV for pH 6.5 and 9, respectively). Evaporation of a thin gold layer shifts the zeta potential to 15 mV and 25 mV.

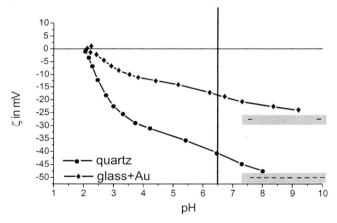

Figure 1. Dependence of zeta potential on pH for glass modified with a thin gold layer, and quartz glass.

First information on the multilayers, built up by the interaction between PD and PSS were obtained by the UV spectroscopy using the absorption band of PSS at 225 nm which is originated from the aromatic ring. As the commercial available branched PD types are known not to be suitable for the preparation of multilayers by the layer-by-layer adsorption technique, this special linear PD was tested for the preparation of multilayers. Figure 2 illustrates the absorbance of multilayers on quartz glass in dependence on the number of double layers. The absorbance is correlated to the

© 2004 WILEY-VCH Verlag GmbH & KGaA, Weinheim

adsorbed amount. The adsorbance for the low molecular mass polyelectrolytes as well as for those with high molecular mass increases with the number of double layers. It may be assumed from the linear rise of the adsorbance that the charge overcompensation in each adsorbed layer was approximately constant from the first layer.

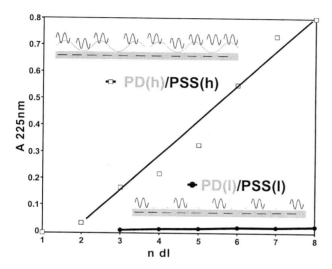

Figure 2.   The absorbance of dried multilayers on quartz glass in dependence on the number of double layers for PD(l)/PSS(l) and PD(h)/PSS(h).

The adsorbed amount per double layer for the system PD(h)/PSS(h) is calculated to be 25×10^{-3} g/m^2, whereas this value for the system PD(l)/PSS(l) is only 0.5×10^{-3} g/m^2, 50 times lower. The molar masses of the polyelectrolytes thus have a big influence on the absolute amount of adsorbed material. However, the linear increase of the adsorbed amount with the number of double layers holds regardless of the molar masses. This linear increase was also found for the system PEI/PSS with the PEI solution adjusted to pH 3.

The SPR method allows to measure the amount of adsorbed material very sensitiv in-situ, taking into account any interaction with the aqueous phase, e.g. swelling. In the next section the system PD/PSS with different molar masses is investigated. The first

© 2004 WILEY-VCH Verlag GmbH & KGaA, Weinheim

steps of the adsorption of a multilayer were followed by a time dependent measurement at which the angle of incidence was fixed to 55.7 and the reflectivity was measured over the time. The angle of incidence was choosen so that the slope of the SPR curve in that angle range is almost linear. According to own model calculations, a small change of the reflectivity is then correlated to the adsorbed amount. The time dependent measurements in Figure 3 show a fast adsorption process for all polyelectrolytes. The adsorption process is finished after some seconds. The rinsing with water does not lead to a desorption and does not prevent the next adsorption step, pointing to a stable multilayer. The comparison of the time dependent measurements of the layers made

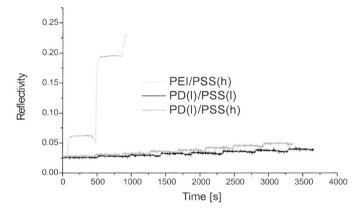

Figure 3.   Time dependent measurements for multilayer systems of PEI/PSS(h), PD(l)/PSS(l) and PD(h)/PSS(h), angle of incidence 55.7°.

from PD(l) and PSS with low and high molar mass shows that the adsorbed amount for PSS with a high molar mass is larger. However, the relation between the both values is much lower than that obtained for the system PD(h)/PSS(h). For both systems the adsorbed amounts for the polycation layers and for the polyanion layers are approximately equally large. The adsorbed amount of PEI in the first layer is much higher than that for PD(l). According to [24] it may be assumed that the adsorption is not only driven by electrostatics, but non-electrostatic forces are responsible for that behavior. This assumption was supported by Monte-Carlo simulations,[31] which shows a large increase of the amount of adsorbed polyelectrolyte chains in the first layer and a higher stability of the total multilayer if a non-electrostatic interaction parameter was

© 2004 WILEY-VCH Verlag GmbH & KGaA, Weinheim

introduced for the polyelectrolyte in the first layer. In our experiments the adsorption of PEI in the first layer leads to a much higher compensation of the surface charge, which than supports an enhanced adsorption of the next layers. The high change in reflectivity in the time dependent measurement due to the adsorption of PSS points to an overcompensation, which is higher again. Since a time dependent measurement with those large reflectivity changes off the linear range cannot be correlated to the adsorbed amount, SPR scans were recorded for more detailed investigations of that system.

The SPR curve with pure water in the measurement cell is shown in Figure 4 a (most left line). The typical values of the best fit to the Fresnel's equations were: dielectic permitivity of the gold layer = -12 + 1.3*i, thickness = 50 nm. These values agree with other data in the literature.[5] Measurements were done after adsorption of each double

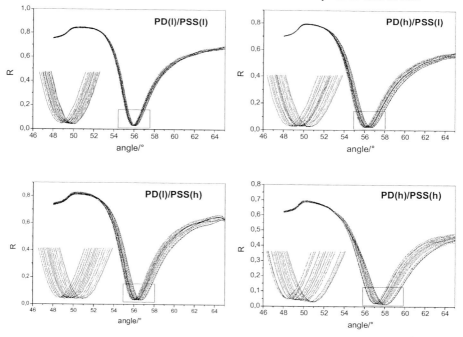

Figure 4. SPR curves measured after the adsorption of multilayers for the systems PD(l) or PD(h) and PSS(l) or PSS(h).
Shift of the SPR minimum to higher angle from 0 to 15 double layers in water.

© 2004 WILEY-VCH Verlag GmbH & KGaA, Weinheim

layer. Each curve in Figure 4 a for the system PD(l)/PSS(l) exhibit a very small but significant shift of the SPR minimum (shown in the inset), whereas the angle of total reflection stayed at the same value. This demonstrates the reliability and sensitivity of the setup and allows for measuring of very small adsorption amounts. The thickness of this double layer was received as a result of the best fit of a four layer model to the data. Since it is not possible to fit the refractive index and the thickness at the same time under these conditions, the refractive index was fixed to be 1.5 for the fit of the thickness. This value is close to that measured for a series of polyelectrolyte multilayers with the Scanning Angle Reflectometry as an independent method. The refractive index determined with that method varied between 1.48 to 1.49.[18] The adsorbed amount is related to the optical parameters according to Feijters equation[33]

$$\Gamma = \frac{d * \Delta n}{\dfrac{dn}{dc}}$$

with thickness d, Δn = n(layer) - n(solution), refractive index n, and the refractive index increment dn/dc with c being the polyelectrolyte concentration of the solution. For the calculation of the adsorbed amount an average value of dn/dc = 0.18 ml*g^{-1} was used.

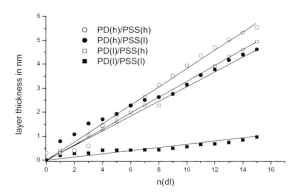

Figure 5.   Layer thickness as a function of the number of double layers for the systems PD(l) or PD(h) and PSS(l) or PSS(h).

© 2004 WILEY-VCH Verlag GmbH & KGaA, Weinheim

The resulting thicknesses, which obviously are averages over the beam spot, are presented as a function of the number of double layers in Figure 5. The multilayer thickness increases almost linearly with the number of double layers, the final thickness after 15 cycles being approximately 1 nm. This gives an averaged thickness per double layer of app. 0.07 nm or an adsorbed amount of $7*10^{-8}$ $g*m^{-2}$. The SPR curves of systems with PD(h) or PSS(h) exhibit larger shifts, see Figure 4 b-d. The resulting average thicknesses and adsorbed amounts per double layer are almost the same for all systems, 0.3 nm and $30*10^{-8}$ $g*m^{-2}$, respectively. The effect of the molar mass of PSS is almost negligible here. It should be remarked that the increase of the multilayer thickness is linear to the number of double layers for all the systems, as found for UV measurements. However, the absolute adsorbed amounts differ considerably between the SPR and the UV measurements. This may be attributed to the different processing conditions (stirring, dry, wet) and substrate surface charges.[34]

The SPR curves for the system PEI/PSS(l) measured after adsorption of each double layer are shown in Figure 6a. On a first glance the differences of the curves are considerably larger compared to the systems with PD as polycation, pointing to larger thicknesses of the multilayers with PEI as a polycation. This agrees with the observation of the reflectivity in Figure 3. At higher number of double layers not only the SPR minimum shifts but also the reflectivity at angles lower than the critical angle is reduced, whereas the critical angle was constant for all curves. This is in agreement with the model at large layer thickness. The fringes at higher angles are artifacts and stems from interferences as the beam spot becomes broad. The curves for the system with PSS(h) show an additional minimum in front of the critical angle, pointing to the exciting of waveguide modes. Exciting of waveguide modes is possible for large layer thicknesses if the layer is smooth and homogeneously. The fitted thicknesses are presented in Figure 7 as a function of the number of double layers. In contrast to the multilayers with PD as polycation, the increase of the layer thickness is not linear but exponential. The effect was not found for the system PEI/PSS on a fixed pH by UV measurements and for a similar system (poly(allylamine*HCl)/PSS(h))[18] although both polycations are weak. The thickness of the multilayer after 10 double layers is approximately 150 nm and 25 nm for PSS(h) and PSS(l) as polyanions, respectively.

© 2004 WILEY-VCH Verlag GmbH & KGaA, Weinheim

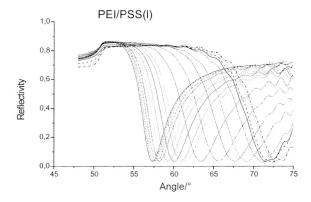

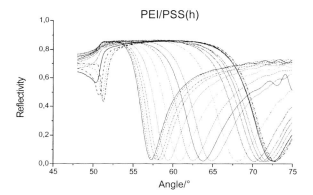

Figure 6.   SPR curves measured after the adsorption of PEI/PSS(l) double layers. Shift of the SPR minimum to higher angle from 0 to 20 double layers in water.

Larger thicknesses could not be fitted satisfactory. The increase of the double layer thickness (adsorbed amount) was approximately 1 nm (10^{-6} g*m^{-2}) at the beginning and 100 nm (10^{-4} g*m^{-2}) for the 15th double layer for the system PEI/PSS(h). The lowest double layer thickness was even more than that in the systems with PD as polycation, even though the zeta potential of the substrate is smaller. The exponential increase of the double layer thickness points to an extended overcompensation of the charge of the

© 2004 WILEY-VCH Verlag GmbH & KGaA, Weinheim

underneath layer by the adsorbed double layer. This extension must increase itself after each double layer, and it may be caused by the change of the pH value from 6.5 to 9 on the change from the PSS to the PEI solution and vice versa. The protonization and, thus, the charge of PEI is higher at pH 6 than at pH 9. For PD and PEI the fraction of charges per repeat unit are approximately 1 and 0.36, respectively, whereas the value for PEI was measured at pH 9, and the value for PD is nearly independent of the pH value. The adsorbed amount may scale to the fraction of charges per chain f and the charge σ of the surface formed by adsorption of the previous layer approximately as $\Gamma \sim \sigma * f^1$.[24] For PSS and PD the absolute σ and f are the same for every adsorption step. According to the model of the built-up of PEL multilayers explained above, half of the charges of every polycation molecule are complexed with the previously adsorbed polyanion, the other half forming the attractive force for the next layer and are exposed to the environment according to the solvent conditions.

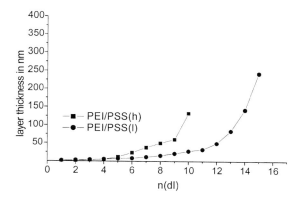

Figure 7.   Fitted thickness as a function of the number of double layers for the systems PEI/PSS(l) and PEI/PSS(h).

In our experiments, however, the conditions are different, particularly due to the change of the polyelectrolyte charge density with the pH of the solution. A qualitativ approximation of the adsorbed amount Γ with the number n of double layers is given in this section. On the adsorption of PEI half of the available charges at pH 9 and the corresponding repeat units interact with the surface with charge σ and leads to an adsorption. Since $f(\text{PEI, pH } 9) = f_A = 0.36$, the adsorbed amount of repeat units is higher, the other repeat units being non-charged at pH 9. On adsorption of the next PSS

© 2004 WILEY-VCH Verlag GmbH & KGaA, Weinheim

layer the pH value has changed to 6.5, resulting in an increase of f of the adsorbed PEI layer to f(PEI, pH 6.5) = f_B = z* f_A with z being app. 1.5. This leads to an increase of the charge of the surface formed by the PEI layer to z*σ, resulting in a high adsorbed amount of PSS, providing the multilayer is still stable. In the first double layer the adsorbed amount may be estimated roughly as

$$\Gamma_1 \sim \frac{\sigma}{f_A} + \frac{\sigma}{f_A},$$

where the left term is for the polycation layer and right term for the polyanion layer. Since f of PSS does not depend on pH the charge of the surface formed by the PSS layer is constant at pH 9, and the adsorbed amount of PEI increases compared to the previous PEI layer. On adsorption of PSS, again, the lower pH leads to a higher surface charge, resulting in a larger amount of adsorbed PSS. The adsorbed amount in the second layer is, thus,

$$\Gamma_2 \sim \frac{\sigma \cdot z_A}{f_A} + \frac{\sigma \cdot z_B}{f_A}.$$

This cycle is repeated on each adsorption step. The adsorbed amount in the ith layer is than, assuming $z_A = z_B$

$$\Gamma_i \sim \frac{2\sigma \cdot z^{i-1}}{f_A}.$$

After repeating the cycle n times the total adsorbed amount is

$$\Gamma_n \sim \frac{2\sigma}{z \cdot f_A \cdot \ln z}(z^n - 1)$$

for z > 1.

According to this rough model to Γ_n, which does not take into account any effects of solvatization, charge screening and non-uniform distribution of the adsorbed material in one layer, lg Γ is correlated to n. The correlations holds also if Γ scales as σ^a *f^b [35] with any a, b. As this was found experimentally for the multilayers with PEI, further detailed investigations will reveal the reasons for the unusual multilayer adsorption more precisely at different pH values and salt concentrations.

© 2004 WILEY-VCH Verlag GmbH & KGaA, Weinheim

Conclusions

The adsorption of PEL multilayers from salt-free solutions was investigated by UV spectroscopy and SPR. These methods ensure a high sensitivity required for the small amounts of adsorbed material. The multilayers were built with PSS with different molar masses as polyanions. As polycations PD with different molar masses and PEI were used. Experiments with the strong polyelectrolytes exhibited a linear increase of the adsorbed amount with the number of double layers. The adsorbed amount per double layer was dependent on the substrate, the adsorption procedure and the molar masses. For the multilayers with PEI as polycation an exponential increase of the adsorbed amount with the number of double layers was found. This behavior was correlated to the charges of PEI at the different pH values of the solutions.

Acknowledgement: The authors thank Norbert Stiehl for a lot of experimental work. This work is funded by the Deutsche Forschungsgemeinschaft.

[1]   G. Decher, *Science* **1997**, *277*, 1232.
[2]   G. Decher, *"Multilayer Films (Polyelectrolytes)"*, In: *The Polymeric Materials Encyclopedia: Synthesis, Properties and Applications,* J. C. Salamone, Ed., CRC Press Inc, Boca Raton 1996, p. 4550.
[3]   P.T. Hammond, *Curr Opin Colloid Interface Sci* **2000**, *4*, 430
[4]   P. Bertrand, A. Jonas, A. Laschewsky, R. Legras, *Macromol. Rapid Commun.* **2000**, *21*, 319.
[5]   C. Pearson, J. Nagel, M. C. Petty, *J. Phys. D: Appl. Phys.* **2001**, *34*, 285.
[6]   T. Hoshi, H. Saiki, S. Kuwazawa, C. Tsuchiya, Q. Chen, J. I. Anzai, *Anal. Chem.* **201**, *73*, 5310.
[7]   A. C. Fou, O. Onitsuka, M. Ferreira, M. F. Rubner, B. R. Hsieh, *J. Appl. Phys.* **1996**, *79*, 7501.
[8]   A. Laschewsky, B. Mayer, E. Wischerhoff, X. Arys, P. Bertrand, A. Delcorte, A. Jonas, *Thin Solid Films* **1996**, *284*, 334.
[9]   D. M. Sullivan, M. L. Bruening, *Chem. Mater.* **2003**, *15*, 281.
[10]   A. A. Antipov, G. B. Sukhorukov, E. Donath, H. Möhwald, *J. Phys. Chem. B* **2001**, *105*, 2281.
[11]   J. Chluba, J. C. Voegel, G. Decher, P. Erbacher, P. Schaaf, J. Ogier, *Biomacromol.* **2001**, *2*, 800.
[12]   G. Decher, *Thin Solid Films* **1994**, *244* (1-2), 772.
[13]   M. Kawaguchi, K. Hayashi, A. Takahashi, *Colloids and Surfaces* **1988**, *31*, 73.
[14]   G. Decher, *Science* **1997**, *277*, 1232.
[15]   P. Bertrand, A. Jonas, A. Laschewsky, R. Legras, *Macromol. Rapid Commun.* **2000**, *21*, 319.
[16]   P. T. Hammond, *Colloid & Interface Science* **2000**, *4*, 430.
[17]   W. Chen, T. J. McCarthy, *Macromolecules* **1997**, *30*, 78.
[18]   G. Ladam, P. Schaad, J. C. Voegel, P. Schaaf, G. Decher, F. Cuisinier, *Langmuir* **2000**, *16*, 1249.
[19]   Ph. Lavalle, C. Gergely, J. C. Voegel, P. Schaaf, G. Decher, F. J. G. Cuisinier, C. Picart, *Macromolecules* **2002**, *35*, 4458.
[20]   E. Kharlampieva, S. A. Sukhishvili, *Langmuir* **2003**, *19*, 1235.
[21]   M. Müller, T. Rieser, K. Lunkwitz, S. Berwald, J. Meier-Haack, D. Jehnichen, *Macromol. Rapid Commun.* **1998**, *19*, 333.
[22]   A. Laschewsky, E. Wischerhoff, M. Kauranen, A. Persoons, *Macromolecules* **1997**, *30*, 8304.
[23]   W. Jaeger, U. Gohlke, M. Hahn, Ch. Wandrey, K. Dietrich, *Acta Polymerica*, **1989**, 40, 161.
[24]   M. Castelnovo, J.-F. Joanny, *Langmuir*; 2000; 16(19); 7524.

© 2004 WILEY-VCH Verlag GmbH & KGaA, Weinheim

[25] Y. Lvov, G. Decher, G.B. Sukhorukov, *Macromol.* **1993**, 26, 5396.

[26] Y. Lvov, G. Decher, H. Möhwald, *Langmuir* **1993**, 9, 481.

[27] R. Advincula, E. Aust, W. Meyer, W. Knoll *Langmuir*, **1996** 12, 3536.

[28] D.G. Kurth, R. Osterhout, *Langmuir*, **1999**, 15, 4842.

[29] J.J. Ramsden, Y.M. Lvov, G. Decher, *Thin Solid Films*, **1995**, 254, 246.

[30] J. Ruths, F. Essler, G. Decher, H. Riegler, *Langmuir*, **2000**, 16, 8871.

[31] R. Messina, Polyelectrolyte Multilayering on a Charged Planar Surface, 2003, in press.

[32] M. Palumbo, C. Pearson, J. Nagel, M. C. Petty, *Sensors and Actuators B* (2003) 90 264-270.

[33]   J.A. de Feijters, J. Benjamins, F.A. Feer, *Biopolymers* **1978**, 7, 1759.

[34]   M. Castelnovo, J.-F. Joanny, *J. Adhes. Sci. Technol.* **1995**, 9, 3, 297.

[35]   I. Borukhov, D. Andelman, H. Orland, *Macromol.* **1998**, 31, 1665.

© 2004 WILEY-VCH Verlag GmbH & KGaA, Weinheim

Multilayers Consisting of Oriented Charged α-Helical Polypeptides

Martin Müller,[1] *Thomas Reihs,*[1] *Bernd Kessler,*[1] *Hans-Jürgen Adler,*[2]

Klaus Lunkwitz[1]

[1] Institute of Polymer Research Dresden e.V. (IPF), Hohe Strasse 6, D-01069 Dresden, Germany
E-mail: mamuller@ipfdd.de

[2] Institute of Macromolecular and Textile Chemistry, Technical University Dresden (TUD), Mommsenstrasse 13, D-01062 Dresden, Germany

Summary: Polyelectrolyte Multilayers (PEMs) consisting of cationic α-helical poly(L-lysine) (PLL) and optionally anionic poly(vinysulfate) (PVS) (i) or α-helical poly(L-glutamic acid) (PLG) (ii) were deposited at substrates texturized by parallel nanoscopic surface grooves, respectively. Using dichroic Attenuated Total Reflexion Fourier Transform Infrared (ATR-FTIR) spectroscopy the consecutive deposition, conformation and macromolecular order of stiff polyelectrolytes within PEMs were studied. From the dichroic ratios of the Amide I and Amide II bands order parameters $S \geq 0.6$ ($S = 1$ for high, $S = 0$ for low order) were obtained suggesting a significant alignment of charged α-helical polypeptides in PEMs. For the PEM consisting of PLL/PVS the deposited amount as well as the order parameter S significantly depended on the molecular weight (contour length) of PLL. Furthermore, the related opening angle γ of a model cone consisting of α-helical PLL rods was proven to be a function of both contour length and width of the confining surface grooves. AFM pictures on PEM-PLL/PVS showed anisotropically oriented worm-like structures As a second system PEMs of PLL and PLG, both in the α-helical conformation, are introduced. A high order parameter of both spectroscopically indistinguishable polypeptides was found. A model of aligned rod/coil (i) and rod/rod (ii) structures was proposed. Finally, multilayers of stiff conductive polymers like polyaniline (PANI) alternating with poly(acrylic acid) (PAC) are introduced. Preliminary results on their deposition and alignment are given.

Keywords: dichroic ATR-FTIR spectroscopy; multilayers; orientation; polyelectrolytes; polypeptides

Introduction

Polyelectrolyte multilayers (PEMs) consecutively adsorbed on solid substrates, which were initiated by Decher,[1] go back to polyelectrolyte complexes (PECs) formed by mixing oppositely charged polyelectrolytes in the solution shown e.g. by Michaels and Kabanov.[2,3] PEMs and PECs have gained much interest in the last decade, since they allow for defined nanoarchitectures and selective surface modification and are challenging objects for polyelectrolyte theory. In the beginning PEMs were claimed to be composed of stratified individual layers. However this model was replaced by another one, according to which commonly used flexible polyelectrolytes cause high entanglement, no distinct layering and a low degree of order in the internal structure.[4, 5] Nevertheless, the generation of PEMs consisting of stratified or lamellar arranged PEL layers is an interesting obective. Macromolecular order in PEMs can be achieved by several concepts: Besides PEMs composed of layered silicates alternating with polyactions[6] (i), those composed of hydrophobic ionenes[7] have been used (ii). As a third concept PEMs composed of charged stiff α-helical polypeptides alternating with oppositely charged strong polyanions or polycations, respectively, were introduced therein[8, 9] (iii) and shall be reviewed and extended here.

Generally, PEMs consisting of charged polypeptides are subjects of fundamental structural studies[8, 9, 10, 11, 12] as well as for biomedical and pharmaceutical applications, which was shown therein.[13, 14, 15, 16] Furthermore, PEMs exposing those polypeptides in both defined conformations and orientations might be a new strategy for biomimetic surface modification as will be shown therein.[17] In that framework PEMs consisting of *in-plane* oriented α-helical polypeptides are expected to bind specifically α-helical rich proteins 'side on', which may be not the case for α-helical polypeptides assembled normal to the substrate or for immobilized polypeptides in other conformations (β-sheet, random coil).

Experimental

Polyelectrolytes

For the consecutive deposition of stiff charged polypeptides alternating with oppositely charged polyelectrolytes (PELs) poly(L-lysine) (PLL, M_w = 3 400, 20 700, 25 700, 57 900, 80 000, 189 400, 205 000, 246 000, 309 500 g/mol, further denoted as 'PLL-3 400 – PLL-309 500', SIGMA-ALDRICH) were combined optionally with poly(vinylsulfate) (PVS, M_w = 162 000 g/mol, POLYSCIENCE) or poly(L-glutamic acid) (PLG, M_w = 70 000 g/mol, SIGMA ALDRICH). All commercial polyelectrolyte samples were used without further purification. PLL, PVS, PLG and PDADMAC were dissolved in Millipore water or in 1 M $NaClO_4$ solution (MERCK, Darmstadt) at PEL concentrations c_{PEL} = 0.01 M. Polyaniline (PANI, PANIPOL, Finland) was dissolved in a mixture of water (pH = 4)/N-methyl-pyrrolidone (NMP) (9:1, v/v) yielding a concentration c_{PANI} = 0.001 M. Poly(acrylic acid) (PAC) was dissolved in water at c_{PAC} = 0.001 M and pH = 4.

© 2004 WILEY-VCH Verlag GmbH & KGaA, Weinheim

Surfaces

The silicon substrates were cleaned[19] and texturized by mechanical treatment[8] as previously reported.

ATR-FTIR spectroscopy

The monitoring of consecutive deposition of the polyelectrolytes and the characterization of the deposited PEMs was performed by in-situ-ATR-FTIR spectroscopy using the SBSR concept[18] to obtain well compensated ATR-FTIR spectra, as it is described therein.[19] Dichroic measurements and data analysis were performed according to a methodology reported therein.[8] IR light was polarized by a wire grid polarizer (SPECAC, UK). The ATR-FTIR attachment was operated on the IFS 55 Equinox spectrometer (BRUKER-Saxonia, Leipzig) equipped with globar source and MCT detector. Typically polycation, rinsing (1 M NaClO$_4$) and polyanion solutions were stepwise injected by syringes in the *in-situ* measuring cell (IPF Dresden, M.M.) each step having a residence time of 15 min. p- and s-polarized ATR-FTIR spectra were recorded after each rinsing step, for which 200 scans were accommodated. To check for undesired time dependent variations the polarized spectra (p, s) were recorded in the sequence ´p - s – p´ and the first p-polarized spectrum was compared with the second one. No spectral differences should appear between the two p-polarized spectra. Either peak intensities or integrated band areas were used for the dichroic ratio determination of the amide bands.

Results and Discussion

In the following deposition and orientation data on the two consecutive polyelectrolyte multilayer (PEM) systems composed of 1. PLL/PVS and 2. PLL/PLG are presented. The influence of parameters like molecular weight (M$_w$), layer number z and of drying will be considered. Briefly deposition results on the PEM consisting of PANI/PAC (3.) are shown.

1. PEM of PLL/PVS

In Figure 1 typical ATR-FTIR spectra on the consecutively adsorbed PEMs (PEM-1 to PEM-5) of PLL/PVS in the presence of 1 M NaClO$_4$ are shown with increasing adsorption steps z from bottom to top. An increasing negative signal of the ν(OH) at about 3 400 cm^{-1} is visible, which is due to the removal of water from the surface upon PEM deposition, which is further commented on therein[20] Additionally, the increasing signals of the Amide I and Amide II

© 2004 WILEY-VCH Verlag GmbH & KGaA, Weinheim

220

band at 1 650 and 1 550 cm^{-1}, due to the peptide groups of PLL, are also related to the PEM growth. The wavenumber positions of both amide bands suggest the α-helical conformation of the PLL component, as it is well known[21] Furthermore, the $v(O=S=O)$ signal around 1 230 cm^{-1} shows also increasing intensity with the adsorption step due to the PVS compound. The courses of these three bands are summarized in Figure 2. As we have shown previously[20] increasing polymer band (e.g. $v(CH)$, $v(C=O)$) intensities A in multilayered systems scale with a function of the type $A(z) = 1 - \exp(-L*z)$ (L: thickness parameter, z: adsorption steps), whereas the increase of the negative $v(OH)$ band scales with a function of the type $A(z) = \exp(L*z) - 1$. A slight indication of a zig-zag-like course is seen in all three curves, which can be explained by a certain removal of the last adsorbed PEL layer by the following one. For example after all PLL steps (odd) in the next step (PVS, even) the PLL amount, as seen by the amide band integrals, drops a little bit down. This is also observed for the water removal monitored by the negative $v(OH)$ band integral. Since in the PLL steps the higher negative $v(OH)$ band increase was observed compared to the previous PVS steps, respectively, PLL seems to form thicker incremental adsorption layers than PVS.

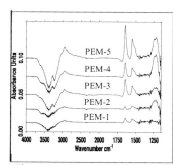 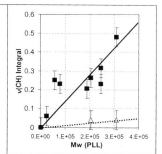

| Figure 1. *in-situ* ATR-FTIR spectra on the consecutive deposition of PLL (205.000 g/mol) and PVS (162.000 g/mol) at Si-crystals (IRE) in the presence of 1 M NaClO$_4$ (unpolarized spectra of PEM-1 to PEM-5). | Figure 2. Deposited PEM amount during consecutive adsorption of PLL/PVS (1 M NaClO$_4$) in dependence of the adsorption step z ratio-nalized by the Amide II, $v(O=S=O)$ and negative $v(OH)$ band (related to Figure 1). | Figure 3. Dependence of the deposited PEM amount on the PLL molecular weight (M_w= 3.400 - 309.000 g/mol) for NaClO$_4$ (full line) and salt free preparation (dotted line). |

In Figure 3 the the deposited amount of PEM-5 of PLL/PVS is shown in dependence of the PLL molecular weight (M_w) for the salt free and the NaClO$_4$ containing system. Qualitatively, the larger adsorbed amount increase of the PEM-5 of PLL/PVS was obtained in

© 2004 WILEY-VCH Verlag GmbH & KGaA, Weinheim

the presence of 1 M NaClO$_4$ compared to the lower one for the salt free system. This is due to both the salt screening effect (i) and the different conformations of PLL (ii). (i) Firstly, PLL in the absence of salt is highly charged, which favors electrostatic attraction to the respective oppositely charged surface in all adsorption steps. However if saturation is reached in every respective adsorption step, further PLL adsorption is prevented by electrostatic repulsion. This is not the case for PLL in the presence of NaClO$_4$, since electrostatic interactions are screened by the high salt concentration (1 M) preventing repulsion. Additionally PVS in the presence of 1 M NaClO$_4$ adopts the more coiled conformation, which might also contribute to the higher PEM adsorbed amount. In this case the PVS coils might act as weak binders between the assembled PLL rods. In how far at that ionic strength electrostatic contributions may play a role at all, has to be treated by theoretical approaches, but can not speculated on here. (ii) Secondly, PLL in the absence of salt is in the random coil conformation and PLL in the presence of NaCLO$_4$ is in the α-helical state and forms a rod. From thermodynamic considerations ($\Delta G = \Delta H - T\Delta S$) rods loose less entropy after adsorption onto surfaces than coils, since the rod conformation does not change ($\Delta S \approx 0$) so much when adsorbed at the surface compared to the conformation of coils loosing degrees of freedom upon surface spreading ($\Delta S < 0$). Furthermore, the enthalpy contribution of rods self-assembling in-plane at the surface (forming a kind of crystal structure) might be higher ($\Delta H < 0$) compared to the enthalpic contribution of coils, which can not show such assembly or cooperative stabilization.

PLL/PVS orientation

Access to the PLL orientation within PEMs can be obtained via dichroic ATR-FTIR spectroscopy, as it is shown in Fig. 4a and 4b. In previous publications[8, 9] we have already shown, that on texturized Si substrates α-helical PLL-205.000 could be oriented along the surface grooves, whereas on untexturized ones this was not the case. Hence texturization was found to be crucial for polymer orientation. Based on these findings texturized substrates are focussed in that report in order to study further parameters of this alignment effect like molecular weight (M$_w$) and layer number (adsorption step) z and additionally PEMs consisting of two α–helical polypeptides. In Figure 4a and Figure 4b p- and s-polarized spectra of the PEM-5 of PLL/PVS in the presence of NaClO$_4$ are shown for PLL with M$_w$ = 25 700 g/mol (further denoted as PLL-25.700, Figure 4a) and 205 000 g/mol (further denoted as PLL-205.000, Figure 4b), respectively. Qualitatively, for PLL-205 000 a high dichroic effect was observed, i.e. the ratios with respect to the Amide I and the Amide II band were

© 2004 WILEY-VCH Verlag GmbH & KGaA, Weinheim

differing very much. This was not the case for PLL-25.700, where the R_y^{ATR} value of the Amide I was about the same of the Amide II. The dichroic ratios are summarized in the Table 1.

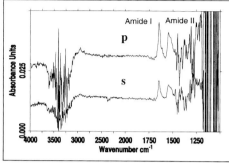 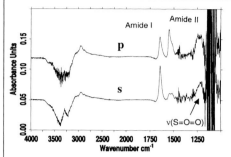

Figure 4a. p- and s-polarized ATR-FTIR spectra of the PEM-5 of PLL (25.700 g/mol) and PVS in the presence of 1 M NaClO₄.

Figure 4b. p- and s-polarized ATR-FTIR spectra of the PEM-5 of PLL (205.000 g/mol) and PVS in the presence of 1 M NaClO₄.

Orientation analysis

For the quantitative analysis of dichroic ATR-IR data a formalism was applied developed by Zbinden and Fringeli[22, 23] at polymer and smaller liquid crystalline compounds and adapted to assembled α-helical polypeptide systems therein,[8, 9] for which the setup is shown in Figure 5a. The basic equation of dichroism[24] is given in the following:

$$A \propto \vec{E}^2 \bullet \vec{M}^2 \cos^2(\vec{E}, \vec{M})\qquad(1)$$

Accordingly, a maximum absorbance A is obtained if the angle between the electric field vector E and the transition dipole moment M is 0° and minimum A is obtained if the angle is 90°. Experimentally, a polarizer with p- and s-setting is commonly used for the generation of polarized light with an electric field vector oscillating parallel or vertical with respect to the substrate normal, respectively. In eq. (2) the experimental dichroic ratio measured in the ATR mode R_y^{ATR} of A_p and A_s, which are the integrated absorbances of a given band measured with p- and with s-polarized light, is given.

$$R_y^{ATR} = \frac{A_p}{A_s}\qquad(2)$$

For simplification R_y^{ATR}, the dichroic ratio measured in ATR mode, can be converted by Equation (3) into a dichroic ratio R^T, which would have been measured in transmission mode, knowing the amplitudes of the relative electric field components (E_x, E_y, E_z) of the evanescent wave (Figure 5b).

© 2004 WILEY-VCH Verlag GmbH & KGaA, Weinheim

$$R^T = R_y^{ATR} \cdot \frac{E_y^2}{(E_x^2 + E_z^2)} \tag{3}$$

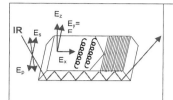

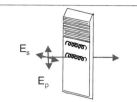

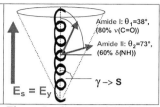

Figure 5a. (From [9], with kind permission of ACS). Experimental setup for dichroic ATR-IR measurements (ATR mode) on texturized (grooves) Si substrates.	Figure 5b. Experimental setup for dichroic transmission IR measurements (transmission mode)	Figure 5c. Cone model of assembled and complexed PLL rods in the surface grooves [9].

Based on a cone model shown in Figure 5c from this R^T value an order parameter S can be calculated knowing the angle θ between the transition dipole moment M and the molecular main axis (e.g. the helical axis for polypeptides) according to Equation (4).

$$S = \frac{(1 - R^T)}{(2R^T + 1)} \cdot \frac{2}{(3\cos^2\theta - 1)} \tag{4}$$

Generally, values of S = 0 are obtained for no order, S = 1 for high order or parallel arrangements and of S = -1/2 for vertical arrangements with respect to a texture or a physical drawing direction, respectively. Since in a macromolecule R^T values can be determined for lots of vibration bands, a molecular model could be built based on their various relative transition dipole moment angles. Finally the above introduced order parameter S is directly related to the cone opening angle γ by Equation (5).

$$\gamma_0 = \arccos\left(\sqrt{\frac{2}{3}S + \frac{1}{3}}\right) \tag{5}$$

Accordingly, high values of γ correspond to low order parameters S or to a low degree of unidirectional alignment of the polymer rods.

For polypeptides the θ values for the Amide I and Amide II band and further for the Amide A band (around 3 300 cm^{-1}) are known from literature[25] according to which $\theta_{AMIDE\ A} = 28°$, $\theta_{AMIDE\ I} = 38°$, $\theta_{AMIDE\ II} = 73°$. The relative location and the angle of the transition dipole moment of the Amide I and Amide II band are depicted in Figure 5c. Hence, to determine the

© 2004 WILEY-VCH Verlag GmbH & KGaA, Weinheim

order parameter of the polypeptide rods three independent informations can be used. The results for the PEM-5 of PLL/PVS are given in Table 1. Generally for IR spectra on PEMs in contact to water we take the order parameter S for the Amide II band as the reference value, since the Amide I band is slightly interfered by the δ(OH) band of water due to incomplete spectral compensation.

Table 1. Dichroic ratios R^T, order parameter S and opening angle γ for PEM-5 of PLL/PVS consisting of PLL with two different M_w.

	PLL-25.700/PVS		PLL-205.000/PVS	
	Amide I	Amide II	Amide I	Amide II
R^T	2.00	1.74	0.60	4.47
S	-0.19	0.10	0.79	0.75
γ	63°	51°	22°	24°

Dependence on the molecular weight M_w

As could be already seen from the dichroic ratios measured for the PLL-25 700 containing PEM-5 and the PLL-205 000 containing PEM-5, whose spectra are shown in Figure 4a and Figure 4b, different orientation level could be qualitatively concluded. Moreover in the Table 1 the order parameters S and cone opening angles γ determined by the equations (2) - (5) are given quantitatively. In Table 1 a significantly higher order parameter (S = 0.75 - 0.79) and thus higher degree of alignment of the PLL rods was obtained for the PLL-205.000 compared to the PLL-25 700 containing PEM (S = -0.19 - 0.10). Obviously the longer PLL-205 000 complexed by PVS, was less flexible in the texturized groove compared to the shorter PLL-25 700, which was able to adopt a less constrained orientation.

Therefore to validate the effect and check for a systematic trend or functional relation further PLL samples with the varying M_w = 3 400, 20 700, 57 900, 80 000, 189 400, 246 000, 309 500 g/mol) were used and deposited consecutively with PVS. Analogously the corresponding order parameters were determined for those PEMs. The result is given in Figure 6a, where the order parameters S are plotted versus the molecular weight (M_w) of all PLL samples. A significant increase of order is seen with increasing M_w. Hence, systematically the longer PLL rods showed the higher alignment in the surface grooves than the smaller ones, which were not influenced by the confined space. For a deeper understanding the theoretically approximated contour lengths L of the different α-helical PLL samples were plotted against the experimentally determined opening angles γ of the rod

© 2004 WILEY-VCH Verlag GmbH & KGaA, Weinheim

assembly, which is shown in Figure 6b. For the calculation of the approximate contour lengths $L_{PLL-3.400}$ - $L_{PLL-309.500}$ the degree of polymerisation (DP) and the known rise per residue of $r = 0.15$ nm for the α-helix[26] were used according to $L = DP * r$.

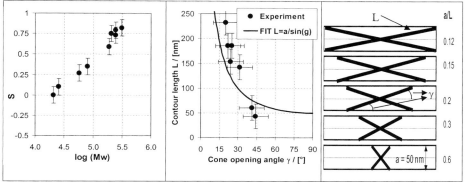

Figure 6a. (partially data from [9] with kind permission of ACS) Dependence of the order parameter S on the molecular weight (M_w) of PLL in PEM-5 consecutively adsorbed with PVS at texturized Si substrates.

Figure 6b. Relation between the contour length L of various PLL samples and the cone opening angle γ due to Equation (5). The analytical function (6) was used.

Figure 6c. Scheme of the confinement effect of a surface groove (width a) on the alignment of stiff poly-mer rods (contour length L). Only two rods of an assumed bundle are shown.

The following simple geometrical model was applied to represent the experimental data,

$$a/L = \sin(\gamma) \tag{6}$$

considering the contour length L as the hypothenusis and a, the groove width, as the opposite cathete of a triangle (Figure 6c). a was the optimised parameter in that approach resulting in a = 50 ± 10 nm. This was quite consistent with the average width of the surface grooves found by AFM.[9] In Figure 6c a scheme on the confinement effect of nanoscopic surface grooves on rod alignment is shown. With these data it can be shown that α-helical PLL form rods in PEMs (i), that their degree of alignment scales with their varying contour length (ii) and that the ratio between groove width a and contour length L is obviously the determining factor for orientation in PEMs of α-PLL/PVS (iii). Arguments for the PLL alignment within surface grooves are currently under discussion with several working groups. Suggestions for main driving forces of PLL rods to confine within the surface grooves and not to exceed the groove space are:[9]

© 2004 WILEY-VCH Verlag GmbH & KGaA, Weinheim

-The higher population of reactive geminal Si-OH groups due to mechanical scratching causes more surface contacts with the α-helical polypeptides, especially PLL compared to the non-groove region.

-The ammonium groups of α-helical PLL rods can form ion pairs not only with the surface charges at the bottom but also with those at the side walls of the groove (half cylinder). Therefore charged rods might form more surface contacts, if they are surrounded by groove walls compared to facing only the planar surface.

Dependence on the layer number

As a further parameter, the influence of the PEL layer number on the unidirectional alignment of PLL rods within the PEM-1 to PEM-5 was studied. As a result in the Figure 7 the order parameter S of PLL-246.800 based on the Amide II band is plotted against the adsorption step $z = 1$ to 5 of PEM-PLL/PVS-z.

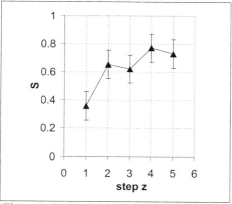

Figure 7. Order parameter S of PLL-246.800 in PEM-z of PLL/PVS in the presence of 1 M $NaClO_4$ in dependence of the layer number z.

An increase from the first PLL layer (PEM-1) from $S = 0.36$ to $S = 0.73$ for the the last PLL layer (PEM-5) was observed and it seemed that PEM-z with $z > 5$ would reach a constant value between $S = 0.7 - 0.8$. This demonstrates that there is a correlation between the number of assemblied PLL rods within the nano grooves and the degree of alignment. Assuming triangular or even conical shape of the surface grooves PEMs might grow from both the bottom and the side walls to fill the empty space. Hence possibly, the more PLL rods are assembled under complexation with the polyanions, the smaller gets the available space in the

© 2004 WILEY-VCH Verlag GmbH & KGaA, Weinheim

grooves, which might enhance the PLL alignment with respect to the groove direction. Furthermore in Figure 7 a modulating course in the order parameter was observed, so that after each polyanion step S is slightly increased compared to the one before. This might be interpreted as a compacting and orienting effect of the outermost polyanion layer onto the underlying PLL rods.

Microscopic characterization (AFM)

In the preceding section it was shown, that the internal structure of PEMs of PLL/PVS exhibited a high degree of orientation induced by parallel surface textures. Hence, it was interesting, how the surface morphology of those thin film assemblies looks like. This was checked by AFM measurements (tapping mode) on the dry PEM sample. In Figure 8a an AFM picture on the PEM-5 of PLL-246.500/PVS, consecutively deposited on a texturized silicon substrate, is given.

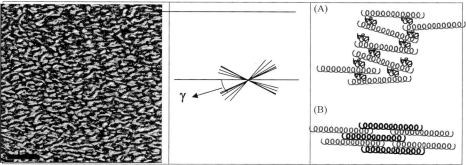

Figure 8a. AFM picture of a PEM-5 composed of PLL-246.500/PVS. The graph scales with 5 x 5 micron (height x width). (texture direction lies parallel to the scaling bar (left), which corresponds to 1 μm).

Figure 8b. Texture vectors of worm-like objects found in Fig. 7a due to oriented nanoscopic complexes of PLL/PVS shifted to the same origin of a double cone with opening angle $\gamma = 30°$.

Figure 8c. (A) Proposed model for PEMs of stiff α-helical PLL and coiled PVS (similar to [9] with kind permission of ACS) (B) Proposed model for PEM consisting of α-helical PLL and α-helical PLG (description in the chapter 2).

© 2004 WILEY-VCH Verlag GmbH & KGaA, Weinheim

A nano-structured surface morphology giving hints for anisotropically oriented elements (white coloured ´worms´) with dimensions of about 200 nm is visible. This might be attributed to partially aligned aggregated PLL/PVS complex moieties. For various of such aligned worm-like objects in the graph texture vectors were drawn, which all had certain inclinations from the texture direction. Recombining them to one center, a double cone could be obtained, which is shown in Figure 8b. Analogously to the orientation model used in FTIR spectroscopy an opening angle of $\gamma = 30°$ of that double cone could be determined, which is in approximate agreement with the opening angle obtained by ATR-FTIR $\gamma = 22°$ for the same PLL-246.500 sample. Although the length scales of these two experiments are different, principles of self similarity might prevail. Hence, the larger scaled surface structures seen by AFM might be related to the lower scaled molecular dimensions, which were probed by IR spectroscopy.

Model

Conclusively in Figure 8c a scheme for the internal structure of PEMs of PLL/PVS is proposed. The layered lammellar structure might be valid in the horizontal as well as in the vertical direction of the PEM. In principle PLL forms the α-helical rods (grey) by specific interaction of the bulky ClO_4^- anions with the polypeptide backbone inserting within the charged ammonium groups of PLL[27] Hence, charged α-helical polypeptides like PLL show an ´anti-polyelectrolyte´ behavior, since they stretch upon salt addition. Whereas, PVS itself shows the classical polyelectrolyte behavior: it forms coils (black) due to the charge screening by the high salt content (1 M $NaClO_4$). Complexing these two PELs supramolecular assemblies are suggested consisting of PLL rods, which are oriented in the case of a texturized (command) surface and PVS coil, which form a kind of glue between the PLL rods.

© 2004 WILEY-VCH Verlag GmbH & KGaA, Weinheim

2. PEM of PLL/PLG

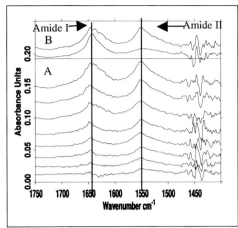 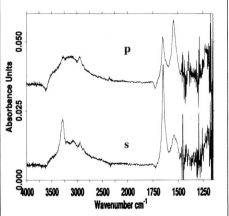

Figure 9a. (A) p-polarized ATR-FTIR spectra on the consecutive deposition of PLL and PLG from NaClO$_4$ solutions in dependence of the adsorption step z = 1 to 8 (from bottom to top). (B) The two uppermost spectra are due to the p- and s-polarized spectrum of PEM-8 of PLL/PLG in the presence of 1 M NaClO$_4$.

Figure 9b. p- and s-polarized ATR-FTIR spectra of the PEM-8 of PLL/PLG in the dry state after rinsing with water and drying by N$_2$ flow.

As it was pointed out above, PEMs of PLL/PVS consist of stiff α-helical PLL and coiled PVS. Consequently, PEMs of two stiff α-helical polypeptides shall be introduced asking on their orientation features. Hence, here we report some preliminary results on the PEM system consisting of PLL and PLG, being both in the α-helical state. In the Figure 9a related *in-situ*-ATR FTIR spectra on the consecutive deposition of PLL/PLG from NaClO$_4$ containing solutions are shown. From the increasing signals of the Amide I band at 1 647 cm^{-1} and of the Amide II band at 1 550 cm^{-1}, the successful built up of PEM-1 to PEM-8 can be proven. Generally for that specific system we only see an overlap of the amide bands from PLL and from PLG. However, since in these spectra the Amide I and Amide II bands did not show significant wavenumber changes upon consecutive adsorption of PLL and PLG, both polypeptides appear to be in the α-helical state. Furthermore, the p- and s-polarized spectra of the PEMs ending either with the PLL or with the PLG component show a high dichroic ratio of $R_Y^{ATR} > 3.0$ with repect to the Amide II band (upper part B) suggesting high order parameters S, which are given in the Table 2.

© 2004 WILEY-VCH Verlag GmbH & KGaA, Weinheim

Additionally the effect of drying was addressed at. For that the spectra of PEM-8 in the dry state are shown in the full range in Figure 9b, in which additionally also the Amide A band (ν(NH)) could be evaluated, since for the dry case this band does not suffer from interferences with the ν(OH) band. The resulting R_y^{ATR} values are summarized in Table 2, whereby for the sake of comparison only the Amide II band was considered.

Table 2. Dichroic ratios R^T, order parameter S and opening angle γ for PEMs of PLL/PLG (based on the Amide II band).

Amide II	PLL-309.500 / PLG-70.000			
	PEM-6, wet	PEM-7, wet	PEM-8, wet	PEM-8, dry
R^T	3.33	4.5	3.63	2.41
S	0.58	0.75	0.63	0.58
γ	32°	24°	30°	32°

From the order parameters $S \geq 0.6$ for the dry as well as the wet state a high unidirectional orientation could be concluded for both the α-helical PLL as well as the α-helical PLG rods, which is based on the average of both spectroscopically undistinguishable polypeptide orientations. A slight increase of alignment was obtained for the PLL terminated PEM (PEM-7) in comparison to the PLG terminated one (PEM-6, PEM-8). Furthermore, drying of the PEM-8 did not change the polypeptide alignment in the wet state, which was also observed for PEMs composed of PLL and a flexible polyanion earlier.[8] Conclusively, based on the high order parameter for both stiff compounds (PLL and PLG) a structural model for such PEMs is given in the Figure 8c (B) suggesting parallel (lammelar) oriented rods of both α-heli-cal PLL and α-helical PLG. Whereas the α-helical conformation of PLL was expected due to the specific PLL/ClO$_4^-$ interaction, this was surprising for PLG, since it is not known that PLG forms the α-helix in 1 M NaClO$_4$ solutions. Hence, it can be speculated if α-helical PLL rods might induce the α-helix of PLG by the exact matching of the opposite charges or by dipole/dipole interaction. Microscopic investigations on this system are currently under way.

3. PEM of PANI/polyanion

Based on the the high macromolecular order observed in the reported PEMs composed of stiff α–helical polypeptides, further synthetic stiff PELs came into consideration asking if they can be oriented in an analogous manner. Interesting macromolecular compounds with this respect

© 2004 WILEY-VCH Verlag GmbH & KGaA, Weinheim

are conductive polymers, since they are charged and expected to be stiff in their doped form. Conductive polymers are commonly known to be poorly soluble in water at pH = 7, to have poor film forming capabilities and to be poorly processable in melts. Hence, to circumvent the first two difficulties, the multilayer technique was applied to immobilize conductive polymers from aqueous solutions, which was initially shown by Rubner and coworkers for polypyrrole (PPY) and polyaniline (PANI).[28, 29, 30]

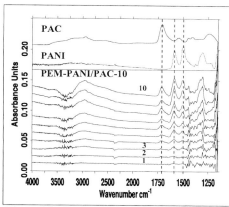 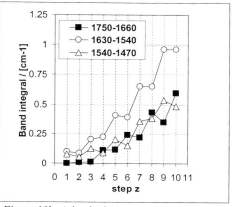

Figure 10a. in-situ ATR-FTIR spectra on the consecutive deposition of PEM-10 consisting of PANI/PAC (from bottom to top: PEM-1 - PEM-10).

Figure 10b. Adsorbed amount versus adsorption step rationalized by the integrated band areas of the ν(C=O) (1750-1660 cm^{-1}) of PAC and the ν(C=C) of the aromatic/chinoidic structures of PANI (1630-1540, 1540-1470 cm^{-1}).

Referring to that studies and in order to exploit our spectroscopic analytical capabilities, doped PANI was consecutively adsorbed in alternating steps with PAC, where acidic PANI solutions of water and NMP (9:1 v/v) were used. As preliminary results in-situ-ATR-FTIR spectra on the consecutive deposition of PANI/PAC at texturized silicon crystals are shown in Figure 10a. From the increasing intensities of IR bands due to PANI and to PAC film formation and growth could be concluded. This is quantitatively shown in Figure 10b, where the integrated band areas of the ν(C=O) of PAC and the ν(C=C) due to the chinoid (1 630-1 540 cm^{-1}) and benzeneoid (1 540-1 470 cm^{-1}) structures of PANI are shown. Interestingly, with every PAC step slight amounts of PANI were removed from the previous PEM. With that we could show that our analytical approach was valid to prove consecutive PANI/PAC deposition.

Moreover, analogously to the polypeptide PEMs we were interested, whether the surface texture had any orientation effect on PANI layers. Up to now applying again dichroic ATR-

© 2004 WILEY-VCH Verlag GmbH & KGaA, Weinheim

FTIR spectroscopy, we could not obtain dichroic effects similar to the polypeptide PEMs. Hence for the moment no alignment of the PANI within the PEM was concluded for that particular system. In future different experimental parameters will be varied to achieve alignment of PANI.

Conclusion

- PEMs consisting α-helical PLL and optionally PVS (i) or α-helical PLG (ii) were deposited at substrates texturized by parallel nanoscopic surface grooves and studied by dichroic ATR-FTIR spectroscopy.

- From the dichroic ratios of the amide bands order parameters $S \geq 0.6$ were obtained suggesting a significant alignment of charged α-helical polypeptides in PEMs containing PLL/PVS (i) as well as PLL/PLG (ii).

- For the PEM consisting of PLL/PVS (i) the deposited amount as well as the order parameter S were significantly dependent on the molecular weight (contour length) of PLL and the opening angle γ of a model cone consisting of α-helical PLL rods was proven to be a function of both contour length and width of the confining surface grooves.

- Models consisting of rods and coils (PLL/PVS) (i) and exclusively of rods (PLL/PLG) (ii), respectively, were concluded from the obtained data.

- AFM pictures on PEM-PLL/PVS showed anisotropically oriented worm-like structures

- Multilayers of conductive PANI alternating with PAC could be successfully deposited. No alignment of PANI could be obtained by the texturized substrate up to now.

Acknowledgement

We thank the Deutsche Forschungsgemeinschaft (DFG) for financial support (SFB 287, B5).

[1] G. Decher, J. D. Hong and J. Schmitt, *Thin Solid Films* **1992**, *210/211*, 831.
[2] A. S. Michaels, R. G. Miekka, *J. Phys. Chem.* **1961**, *65* (10), 1765.
[3] V. A. Kabanov, *Polymer Science*, **1994**, *36*, 2.
[4] G. Decher, *Science* **1997**, 1232.
[5] M. Castelnovo, J. F. Joanny, *Langmuir* **2000**, *16*(19), 7524.
[6] E. R. Kleinfeld, G. S. Ferguson, *Science* **1994**, 265, 370.
[7] X. Arys, A. Laschewsky, A.M. Jonas, *Macromolecules* **2001**, *34*, 3318.
[8] M. Müller, *Biomacromolecules* **2001**, 2(1), 262.
[9] M. Müller, B. Kessler, K. Lunkwitz, *J. Phys Chem. B* **2003**, *107* (in press).
[10] C. Picart, P. Lavalle, P. Hubert, F. J. G. Cuisinier, G. Decher, P. Schaaf, J. C. Voegel, *Langmuir* **2001**, *17*(23), 7414.
[11] P. Lavalle, C. Gergely, F. J. G. Cuisinier, G. Decher, P. Schaaf, J. C. Voegel, C. Picart, *Macromolecules* **2002**, *35*(11), 4458.

© 2004 WILEY-VCH Verlag GmbH & KGaA, Weinheim

[12] F. Boulmedais, P. Schwinte, C. Gergely, J.C. Voegel, P. Schaaf, *Langmuir* **2002**, *18*(11), 4523

[13] P. Schwinte, J.C. Voegel, C. Picart, Y. Haikel, P. Schaaf, B. Szalontai, *J. Phys. Chem. B* **2001**, *105*(47), 11906

[14] J. Chluba, J.C. Voegel, G. Decher, P. Erbacher, P. Schaaf, J. Ogier, *Biomacromolecules* **2001**, *2*(3), 800

[15] P. Schwinte, V. Ball, B. Szalontai, Y. Haikel, J.C. Voegel, P. Schaaf, *Biomacromolecules* **2002**, *3*(6), 1135

[16] L. Richert, P. Lavalle, D. Vautier, B. Senger, J.F. Stoltz, P. Schaaf, J.C. Voegel, C. Picart, *Biomacromolecules* **2002**, *3*(6), 1170

[17] M. Müller (in preparation)

[18] U.P. Fringeli, in *'Encyclopedia of Spectroscopy and Spectrometry'*, J.C. Lindon, G.E. Tranter, J.L. Holmes (eds), Academic Press, 2000

[19] M. Müller, T. Rieser, K. Lunkwitz, S. Berwald, J. Meier-Haack, D. Jehnichen, *Macromol. Rapid Commun.* **1998**, *19*(7), 333

[20] M. Müller, in *'Handbook of Polyelectrolytes and Their Applications'*, Eds. S.K. Tripathy, J. Kumar, H. S. Nalwa, Vol. 1, American Scientific Publishers (ASP), 2002, pp. 293-312

[21] N.A. Nevskaya and Y.N. Chirgadze, *Biopolymers* **1976**, *15*, 637

[22] R. Zbinden, *IR-Spectroscopy of High Polymers*, Academic Press, NY 1964

[23] U.P. Fringeli, M. Schadt, P. Rihak, Hs. H. Günthard, *Z. Naturforsch.* **1976**, *31a*, 1098

[24] J. Michl and E.W. Thulstrup, Spectroscopy with polarized light, VCH Publishers, New York 1986

[25] T. Miyazawa, *J. Chem. Phys.*, 32, 1647 (1960) &. S. Krimm, *J. Mol. Biol.* **1962**, *4*, 528

[26] G.E. Schulz, R.H. Schirmer, Principles of protein structure, Springer, N.Y. 1985

[27] G. Ebert und Y.-H. Kim, *Progr. Colloid & Polymer Sci.*, **1983**, *68*, 113

[28] A.C. Fou and M.F. Rubner, *Macromolecules* **1995**, *28*, 7115

[29] J.H. Cheung, M*acromolecules* **1997**, *30*, 2712

[30] W.B. Stockton and M.F. Rubner, *Macromolecules* **1997**, *30*, 2717

© 2004 WILEY-VCH Verlag GmbH & KGaA, Weinheim

RETURN TO: CHEMISTRY LIBRARY
100 Hildebrand Hall • 510-642-3753

LOAN PERIOD	1	2	3
4		5	6

2-HR USE.

ALL BOOKS MAY BE RECALLED AFTER 7 DAYS.
Renewals may be requested by phone or, using GLADIS,
type **inv** followed by your patron ID number.

DUE AS STAMPED BELOW.

FORM NO. DD 10
3M 5-04

UNIVERSITY OF CALIFORNIA, BERKELEY
Berkeley, California 94720–6000